Communications in Computer and Information Science 387

T0224111

For further volumes:
http://www.springer.com/series/7899

Shuigeng Zhou · Zhiang Wu (Eds.)

Social Media Retrieval and Mining

ADMA 2012 Workshops, SNAM 2012
and SMR 2012
Nanjing, China, December 15–18, 2012
Revised Selected Papers

 Springer

Editors
Shuigeng Zhou
Fudan University School of Computer
  Science
Shanghai
People's Republic of China

Zhiang Wu
University of Finance and Economics
Nanjing
People's Republic of China

ISSN 1865-0929   ISSN 1865-0937 (electronic)
ISBN 978-3-642-41628-6   ISBN 978-3-642-41629-3   (eBook)
DOI 10.1007/978-3-642-41629-3
Springer Heidelberg New York Dordrecht London

Library of Congress Control Number: 2013953641

Printed on acid-free paper

Springer is part of Springer Science+Business Media (www.springer.com)

Preface

The all-round infiltration of Web 2.0 these days substantially promotes the development of social networks that are reshaping the way people and organizations interact. Against this background, two workshops on social media retrieval and mining were held in conjunction with the 8th International Conference on Advanced Data Mining and Applications (ADMA 2012).

ADMA 2012 was held in Nanjing, China, December 15–18, 2012, and the workshops were held on December 15. The ADMA 2012 program featured five keynote speeches from prestigious scientists in data mining and related areas, including the 2010 Turing Award winner, Prof. Leslie Valiant from Harvard University, Prof. Bo Zhang from Tsinghua University, Prof. David Bell from Queen's University Belfast, Prof. Qiang Yang from Hong Kong University of Science and Technology, and Prof. Ester Martin from Simon Fraser University. Over 150 people from more than 20 countries and areas attended the conference and the workshops.

This separate CCIS volume contains 15 selected papers from the ADMA 2012 workshops, including the International Workshop on Social Network Analysis and Mining (SNAM 2012) organized by Dr. Cuiping Li and Dr. Gao Cong and the InternationalWorkshop on Social Media Mining, Retrieval and Recommendation Technologies (SMR 2012) organized by Dr. Guandong Xu and Dr. Bo Jiang.

We would like to thank the ADMA 2012 workshop co-chairs, Dr. Hong Cheng from The Chinese University of Hong Kong and Dr. Jianyong Wang from Tsinghua University, for their careful evaluation and selection on the workshop proposals. We thank all workshop organizers for their efforts in organizing two successful workshops. We also thank the authors for their submissions and the workshop Program Committee members for their excellent reviewing work. We hope that the papers in the proceedings are interesting and stimulating.

August 2013
Shuigeng Zhou
Zhiang Wu

Organization

International Steering Committee

Xue Li, Chair	University of Queensland, Australia
Kyu-Young Whang	Korea Advanced Institute of Science and Technology, Korea
Chengqi Zhang	University of Technology, Sydney, Australia
Osmar Zaiane	University of Alberta, Canada
Qiang Yang	The Hong Kong University of Science and Technology, SAR China
Jie Tang	Tsinghua University, China

Honorary Chair

Ruqian Lu	Chinese Academy of Sciences, China

General Co-chairs

Chengqi Zhang	University of Technology, Sydney, Australia
Zhi-Hua Zhou	Nanjing University, China

Program Co-chairs

Shuigeng Zhou	Fudan University, China
Songmao Zhang	Chinese Academy of Sciences, China
George Karypis	University of Minnesota, USA

Workshop Co-chairs

Hong Cheng	The Chinese University of Hong Kong, China
Jianyong Wang	Tsinghua University, China

Tutorial Co-chairs

Joshua Zhexue Huang	Shenzhen Institutes of Advanced Technology, Chinese Academy of Sciences, China
Xingquan Zhu	University of Technology, Sydney, Australia

Publicity Co-chairs

Aixin Sun   Nanyang Technological University, Singapore
Bin Wang   Nanjing University of Finance and Economics, China

Local Co-chairs

Jie Cao   Nanjing University of Finance and Economics, China
Ruibo Yao   Focus Technologies Co. Ltd., China

Co-organized by

Nanjing University of Finance and Economics, China
Focus Technologies Co. Ltd., China

Sponsored by

University of Technology, Sydney, Australia
Chinese Academy of Sciences, China
Fudan University, China
Nanjing University, China

Contents

Social Image Retrieval and Visualization

Networks and Graphs
Processing

Towards Graph Summary and Aggregation: A Survey

Jinguo You[1(✉)], Qiuping Pan[1], Wei Shi[2], Zhipeng Zhang[1],
and Jianhua Hu[1]

[1]School of Information Engineering and Automation,
Kunming University of Science and Technology, Kunming 650500 Yunnan, China
jgyou@126.com
[2]Department of Computer, Xidian University, Xi'an 710071 Shanxi, China

Abstract. To obtain the insight in a single large graph and to save the space consumption for graph mining, the graph summary transforms the input graph into an aggregated concise super-graph represented by supernodes and super-edges. In this paper, we investigate current algorithms of the graph summary and aggregation. We provide the classification of them in terms of partition criterion or information lossless. Further, the main graph summary algorithms are compared and discussed in detail. In the end, we give the challenges and future works.

Keywords: Graphs · Networks · Summarization · Aggregation

1 Introduction

Recently graph-structured data [1–3] is commonly collected and analyzed in a wide spectrum of application domains such as the Web, social networks, communication networks, biological networks etc.

A common scenario in all of the above applications is the need to analyze large graphs with millions or even billions of nodes and edges. It is very difficult to obtain the full insight by mere visual inspection due to the complex structure and the increasing the size of graphs. On the other hand, to develop graph mining algorithm that can scale to gigantic size graphs is a non-trivial challenge, especially when the graph size is far beyond main memory. Moreover, notice that recently another interesting problem is that social networks need to prevent particular attacks to achieve privacy and anonymity protection for social networks.

Graph summary and aggregation is a potential solution to these above problems. The goal of graph summarization is to obtain a concise graph representation constructed from a single large graph. Intuitively, a graph summary is a much higher level abstraction that removes some trivial details from the input graphs. In the summarized graph, each node corresponds to a group of nodes in the input graph, and each edge represents the edges between all pair of connecting nodes from the two groups respectively. It provides an insight into the coarse-level structure of the graph, the main groupings, and the important relationships between the groups [4].

S. Zhou and Z. Wu (Eds.): ADMA 2012 Workshops, CCIS 387, pp. 3–12, 2013.
DOI: 10.1007/978-3-642-41629-3_1, © Springer-Verlag Berlin Heidelberg 2013

There maybe some different technical terminologies that are close to or the same with graph summary in literatures, i.e. graph compression [5–8], graph synopsis [1], graph simplification [9], network abstract [10] etc. For convenience, they are uniformly called graph summary/aggregation in this paper. Also we emphasize that the result should be a graph and the input graph is one single graph. The most extensive literature exists in the field of graph compression, especially for Web graph [5–8]. But much of the work computes compressed representation that is not really a graph. Therefore the intuition into the structure information is lost. Raghavan and Garcia-Molina [8] proposed a novel Web graph representation, called S-Node. It contains two-level representation components, one is a super node graph, the other is a set of low level graphs, such as intranode graphs and superedge graphs, which describe the input graph structure information intra a supernode and between supernodes and are highly compressible by using referencing encoding. Recently great interests arise in graph summary and aggregation. Navlakha et al. [4] proposed lossless and lossy generic graph compression that consists of two parts: a graph summary and a set of edge corrections. In some sense, it is similar to S-Node above, but it is based on sound information theoretic underpinnings. The information theory is also adopted in [11–14]. Tian et al., Tian and Patel [15, 16] proposed multiple resolution OLAP-style graph summaries called k-SNAP based on user-selected node attributes and relationships. The summarization combining with OLAP can also be found in graph OLAP [17–20] and graph cube [21]. Zhang et al. [22] generalized k-SNAP from categorical node attributes to numerical node attributes, while [9] mainly from unweighted graph to weighted graph. Rodrigues et al. [23] presented supergraph with hierarchies to visualize large graphs in their GMine system.

In this paper, we investigate the current related graph summary and aggregation algorithms extensively. We provide the classification of these algorithms. Further, the main graph summary algorithms are compared and discussed in detail. We give the challenges and future works in the end.

2 Preliminary

Assume that an input graph $G(V,E)$ is simple, undirected, and unweighted; V denotes the set of n nodes/vertices $V = \{v_1,..., v_n\}$ and E denotes the set of edges among these nodes.

An attributed graph is denoted as $G(V,E,A)$, where A is the set of m attributes associated with nodes in V for describing vertex properties. Each vertex $v_i \in V$ is associated with an attribute vector $[a_1(v_i), ..., a_m(v_i)]$ where $a_j(v_i)$ is the attribute value of vertex v_i on attribute a_j.

Given an input graph $G(V,E)$, a summary $G(V_S, E_S)$ is an aggregated graph structure in which each node $v \in V_S$, called a supernode, corresponds to a set A_v of nodes in G, and each edge $(u, v) \in E_S$, called a superedge, represents the edges between all pair of nodes in A_u and A_v. For example, consider the input graph in Fig. 1a. In this case, the nodes are uniquely identified with integer values. Figure 1b presents a possible partition of the nodes of the input graph. Figure 1c shows the corresponding summary that consists of two supernodes representing the set

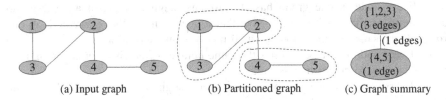

Fig. 1. Graph summarization example

Table 1. Comprison between graph summary and graph clustering

	Grouped nodes connectivity	Dense/sparse	Global/local	The result is a graph?
Graph clustering	Yes	Dense	Local	No
Graph summary	Not mandatory	Dense/sparse	Global	Yes

{1,2,3}and the set {4,5} of nodes in the input graph respectively. The weight on the superedge denotes the total number of edges between the connecting nodes belonging to the two group sets respectively.

Note that graph summary and graph clustering have some similarities. Essentially, they all partition or group similar nodes in a graph into a cluster or a supernode. But there are differences between them. Graph clustering is focused on finding the graph density structure which doesn't need to represent the whole graph. In contrast, graph summarization is focused on preserving the most nodes and edges' statistics of the underneath original graph and constructing a high level succinct super graph.

The following Table 1 gives further comparison between graph summary and graph clustering. The grouped nodes in graph clustering should be densely connected, whereas it is not mandatory in graph summary. Moreover, graph summary considers both dense and sparse structure of the input graph from a global view.

3 Classification

Homogeneous partition/group is an important conception in the graph summary. First we consider the general partition that randomly partitions nodes into groups which are not necessarily attribute homogeneous or structural cohesive. Namely in random partitioning groups, the nodes in the same group need not necessarily have the neighbor relation with nodes in another group or the nodes in the same group need not necessarily have similar attribute values. The random partition is a generalized partition and represents the universal set u for diverse partition criteria.

Next we give the homogeneous partition criteria as following:

i. Nodes in the same group have the same attribute values according to attribute information.

ii. Every node in one group has at least one neighbor node in another group according to neighbor information. Note that it doesn't mean that every node in one group has the same number of edges with nodes in another group.

iii. Every node in one group has the same number of neighbor nodes in another group according to connection strength.

iv. Every node in one group has the same neighbor nodes in another group according to reconstruct ability. Criterion *iv.* means that the original graph can be reconstructed from the super graph without introduced errors.

Criterion *i., ii., iii., iv* represent the homogeneous information from the attribute, neighborhood, connection and reconstruction view respectively. When Criterion *i.* is satisfied, we obtain the attribute homogeneous groups. Criterion *ii., iii., iv.* are based on the neighbor relationship between one group's nodes and another group's nodes.

The following Fig. 2 shows the example of diverse homogeneous partition criteria. There are five groups (i.e. A_1, A_2, A_3, B, and C) and the relationships between these groups in Fig. 2. All the groups satisfy the basic criterion *i*. Further, the neighborhood between group A_1 and group B meets criterion *ii.*, since every blue node in group A_1 relates to white nodes in group B. The neighborhood between group A_2 and group B meets criterion *iii.*, for every red node in group A_2 has the 2 neighbor white nodes in group B. The neighborhood between group A_2 and group C meets criterion *iv.*, for every red node in group A_2 has the same neighbor white nodes in group C.

In terms of the relationships among these criteria, we have the following Fig. 3. Criterion *ii., iii., iv.* asymptotically approach the structure cohesiveness in turn. The bigger the criterion's number is, the stricter it is. That means Criterion *iv.* is the most stricter criterion. When Criterion *iv.* is satisfied, then Criterion *ii., iii.* should be satisfied. Therefore, we say that Criterion *iv.* is contained by Criterion *iii.* and Criterion *iii.* is contained by Criterion *ii.* The intersections of *i.-ii., i.-iii.* and *i.-iv.* all belong to the hybrid graph summary that achieves both attribute homogeneity and structure cohesiveness.

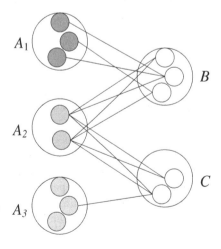

Fig. 2. Graph partition criteria example

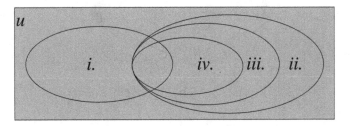

Fig. 3. Relationships among graph partition criteria

Table 2. Classification of current algorithms

	u	*ii.*	*iii.*	*iv.*
u	GraSS [11]			MDL-based [4] Weighted Graph [9]
i.	Graph OLAP [17–20], Graph Cube [21]	SNAP/K-SNAP [15, 16, 22]	Entropy-based [12–14]	

From the above Fig. 3, we obtain 8 combination kinds for diverse graph summary algorithms (see the following Table 2).

Although the philosophy in GraSS [11] can deal with the summaries based on random partitioning, [11] in fact utilizes the structure information to measure the quality of summaries. So we treat GraSS as the structure-based graph summary in the next section for convenience. The graph OLAP [17–20] and graph cube [21] methods in criterion *i.* construct multiple summaries at diverse resolution views. That means they partition the nodes in terms of not only all base attributes, but also the combinations of all attributes as multidimensional spaces. Besides criterion *ii.*, SNAP [15, 16] considers the typed edges, which means the edge between the connecting nodes from the two groups respectively should have the same relation types. K-SNAP [15, 16] reasonably relaxes the edge cohesiveness requirement to improve summaries. Navlakha et al. [4] first introduces the MDL (Minimum Description Length) principle to compute graph summaries. It tends to utilizes the criterion *iv.* to construct super graph. If the criterion *iv.* is not satisfied, then it adds corrections to adjust the summary.

From a high level view, we have three main kinds of graph summaries: attribute-based, structure-based and hybrid of the both. The attribute-based summary meets the criterion *i.* It groups the graph by the vertex attribute similarity in an attributed graph so that nodes within one group have similar attribute values, while nodes between groups have quite different attribute values. On the contrary, the structure-based summary meets at least the criterion *ii.* It partitions the graph by the structure cohesiveness so that nodes within one group are close to each other in terms of structure, while nodes between groups are distant from each other.

The attribute-based groups have a rather structurally loose intra-group, while the structure-based groups generated have a rather random distribution of vertex attributes

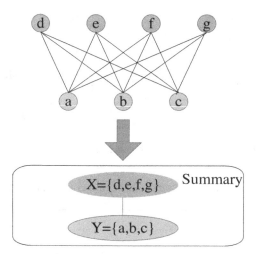

Fig. 4. Bipartite graph summarization

within groups. Because of the two disadvantages, the hybrid of both attribute-based and structure-based methods is proposed recently. When partitioning the input graph, it considers both the attribute similarity and the structure cohesiveness.

Note that the structure-based graph summary commonly utilizes the special structure such as bipartite subgraphs and cliques that are easy to be compressed and aggregated (see Fig. 4). The intuition derives from the well-established observed properties that Web graphs and social networks etc. in the reality world frequently contain nodes that are densely inter-linked with one another or have nodes that have the same (or very similar) neighborhoods. In such scenario, we can collapse such nodes into a supernode and add a single superedge between two supernodes.

From another point of view, the graph summary can be classified into lossless summary and lossy summary. The former means that the graph summaries can be reconstructed to the original graph or can exactly answer the queries derived from the original graph. On the contrary, the latter looses the lossless summary condition and represents the input graph approximately, especially with bounded errors. Thus it reduces the size of the summarization efficiently and still gives reasonable answers to the data analysis needs.

In the following section, we describe the main current algorithms in detail and give some comparisons and discussions.

3.1   Attribute-Based Graph Summary

Graph OLAP and Cube. OLAP has been widely used in data warehouses for supporting decision making. The data cube, as an OLAP's model, computes every combination of analysis dimensions (i.e. group by attributes) to improve online analytical performance. To facilitating OLAP and decision making in attributed networks, graph OLAP and graph cube are studied in [17–21]. Chen et al., Li et al., Qu et al.,

Li et al. [17–20] proposed two kinds of graph OLAP: Informational OLAP and Topological OLAP depending on whether the node type is changed in coarser level or fine level OLAP multidimensional spaces/views.

Like grouping the tuples and summarizing one group into one aggregated tuple in the relation database, graph cube partitions the graph nodes in terms of group by attributes associated with nodes. The difference is that the aggregate results in networks are graphs, namely aggregate networks, in coarser levels of granularity within different multidimensional spaces. Further, Zhao et al. [21] classifies graph cube query kinds into cuboid queries and crosscuboid queries. The implementation of graph cube is also proposed in [21] based on heuristic rules such as selecting the k cuboids with the highest benefit or just computing from minimal level of the cube lattice, which exists in well-studied data cube techniques.

Although Zhao et al. [21] proposed the graph cube model which is of importance, some open problems are still left. The current graph cube model mainly focuses on grouping by the nodes in the graph only in terms of node attributes, resulting in the structure information lost. Cubing and aggregating the network not only based on node attributes but also based on structural information will be an interesting and promising topic. Besides, due to the complexity of the network, compressing the multiple aggregate networks in different multidimensional spaces is more necessary and more difficult than traditional OLAP.

3.2   Structure-Based Graph Summary

GraSS. Not like other methods, GraSS [11] infers the summary model and method from the result graph summary. Reconstructing the input graph from the result summary graph introduces uncertainty. There are diverse possible worlds of the input graph under the summary, which can be represented by the expect adjacency matrix. For random partitioned groups, GraSS assesses the partition quality by the reconstruction error, i.e. the difference between the summary graph's expect adjacency matrix and the input graph's expect adjacency matrix. GraSS finds the graph summary with minimum reconstructing error.

MDL-Based Graph Summarization. The MDL-based representation [4] of a given graph G consisting of a graph summary S and a set of corrections C, denoted by $R(G) = (S, C)$. The graph summary S produces a high-level compact aggregated graph, while the corrections portion C specifies the list of edge-corrections that are applied to the summary to recreate G.

Based on MDL principle, the graph summary S and the corrections C essentially describe the input graph. Thus the best representation of G is to minimize the cost of $S + C$. To obtain better compression, approximate representation is proposed which reconstructs the graph with bounded error (ε). Navlakha et al. [4] presents the cost reduction measure $s(u, v)$ for any given pair of nodes (u, v), which counts the benefit if the two pair of nodes are merged. Then a greedy algorithm is presented to pick the pair with maximum reduction at every step until no positive reduction exists. A random algorithm is also presented which randomly chooses a node u and then merges the best pair containing u.

3.3   Hybrid Graph Summary

SNAP and K-SNAP. To allow users to choose their interesting attributes and relationships in networks to produce a small and intuitive summaries, Tian et al. and Tian and Patel [15, 16] presented SNAP operation and K-SNAP operation. SNAP (Summarization by Grouping Nodes on Attributes and Pairwise Relationships) constructs a summary graph by grouping nodes based on customized node attributes and relationships of the input graph. That means that i) nodes in the same group have the identical values for all attributes A associated with the nodes; ii) if there are relationships R between the two groups Φ_1, Φ_2, then every node in group Φ_1 will find the relevant nodes in group Φ_2 for all the relationships in R. When the above two conditions are satisfied, a (A, R)-compatible grouping is obtained. SNAP produces the global maximum (A, R)-compatible grouping. In other words, SNAP ensures a summary with the minimum groups.

In the reality, however, there are noisy and uncertain relationships frequently for most graph data, SNAP on these data may produce a large number of small groups, resulting a large size and even meaningless summary. Thus K-SNAP is proposed which relaxes the homogeneity requirement for the relationship and still maintains the consistency in attributes. By providing value k to control the number of groups, K-SNAP allows users to drill-down or roll-up the summaries in an OLAP-style way. Because there may be different grouping strategies for size k groups compatible with attributes given k, Zhou [10] introduces Δ to assess the quality of the grouping strategies. Δ counts the minimum number of differences in relationship participations between two groups. Thus the smaller the value of Δ is, the better the quality of the grouping. K-SNAP produces the k size attributes compatible groups with the minimum Δ, which is proved to be a NP-complete problem.

Entropy-Based Graph Summarization. The entropy-based graph summary method [13] presents a unified entropy model that unified both attribute information and structure information. According to information theory, entropy is a measure of the uncertainty associated with a random variable. The more similar the nodes in a group are, the smaller the entropy is. Liu et al. [13] proposed the approximate homogeneous partition that relaxes the criterion $i., ii., iii.$ to achieve more compact graph summaries.

To find the optimal approximate homogeneous partition, two methods are introduced. The first one is the agglomerative algorithm, which first finds the exact homogeneous partition that satisfies the criterion $i., ii., iii.$ Then it calculates the initial value of total weighted entropy of the exact partition after merging each possible node pair. Finally the approximate homogeneous partition is obtained. The main idea of the algorithm is to maintain a matrix to record the change in total weighted entropy for each pair of groups if they are merged, and merge the pair of groups with the minimum change in total weighted entropy repeatedly.

The second method is called the divisive k-means algorithm using a greedy strategy. It starts from one node set, which is the input graph. First, it splits a random perturbation of nodes in the groups with the maximum weighted entropy. Then, the nodes are checked following the order in the perturbation. The method only moves the node which from the old node set to a new node set decreases the total weighted entropy.

4 Conclusions and Future Works

In this paper, we investigate the current algorithms about graph summary and aggregation extensively. We give the classification of them, i.e. the three main kinds: attribute-based, structure-based and hybrid and the five homogeneous partition criteria. In detail, we give five homogeneous partition criteria on how to group the nodes of the input graph. We further review the current main graph summary and aggregation algorithms.

With the increasing size of large graphs and more diverse applications modeled as graphs, new challenges have arose in the graph summary and visualization field.

Scalable Graph Summarization. Graph summarization is to solve scalable graph mining in massive graphs, but developing highly scalable graph summary algorithms is still an interesting problem. Most current works assume memory-resident graph summary algorithms. Therefore, it is worthy to develop efficient disk-resident algorithms or employ the parallel programming pattern such as MapReduce.

Various Type Graph Summarization. For some applications, the edge/link between nodes are uncertain because of the noise etc. To address this problem, we need probabilistic graph summarization which is little done at present. In some settings, both attribute and structure information should be considered at the same time when partitioning the given input graph. In spite of a little works, graph summary based on the hybrid of attribute and structural similarities still warrants a thorough study due to its importance. Thus far, we only consider summarizing one single input graph in this paper. How to summary multiple graphs into a graph is another interesting area of future research.

Acknowledgments. This work is supported by the Natural Science Foundation of Yunnan Province, China (2010ZC030) and is partially done when the author(s) visited Sa-Shixuan International Research Centre for Big Data Management and Analytics hosted in Renmin University of China. This Center is partially funded by a Chinese National "111" Project "Attracting International Talents in Data Engineering and Knowledge Engineering Research".

References

1. Aggarwal, C., Wang, H.: Managing and Mining Graph Data. Springer, New York (2010)
2. Chakrabarti, D., Faloutsos, C.: Graph mining: laws, generators, and algorithms. ACM Comput. Surv. **38**(1), 2 (2006)
3. Leskovec, J., Kleinberg, J., Faloutsos, C.: Graphs over time: densification laws, shrinking diameters and possible explanations. KDD'05: Proceedings of the 11th ACM SIGKDD, pp. 177–187. ACM, New York (2005)
4. S. Navlakha, R. Rastogi, and N. Shrivastava. Graph summarization with bounded error. In: Proceedings of the 2008 ACM-SIGMOD International Conference Management of Data (SIGMOD'08), Vancouver, Canada, pp. 419–432, June 2008
5. Adler, M., Mitzenmacher, M: Towards compressing web graphs. In: Data Compression Conference, pp. 203–212 (2001)

6. Boldi, P., Vigna, S.: The webgraph framework i: Compression techniques. In: WWW, pp. 595–602 (2004)

7. Suel, T., Yuan, J.: Compressing the graph structure of the web. In: Data Compression Conference, pp. 213–222 (2001)

8. Raghavan, S., Garcia-Molina, H.: Representing the webgraphs. In: ICDE, pp. 405–416 (2003)

9. Toivonen, H., et al.: Compression of weighted graphs. In: Proceedings of the 17th ACM SIGKDD International Conference on Knowledge Discovery and Data Mining. ACM (2011)

10. Zhou, F.: Methods for network abstraction. University of Helsinki, Helsinki (2012)

11. LeFevre, K., Terzi, E.: GraSS: graph structure summarization. In: SDM 2010, pp. 454–465 (2010)

12. Liu, Z., Yu, J.X.: On summarizing graph homogeneously. In: Database Systems for Advanced Applications, pp. 299–310 (2011)

13. Liu, Z., Yu, J.X., Cheng, H.: Approximate homogeneous graph summarization. JIP **20**(1), 77–88 (2011)

14. Yin, D., Gao, H., Zou, Z.: A novel efficient graph aggregation algorithm. J. Comput. Res. Devel. **48**(10) (2011)

15. Tian, Y., Hankins, R.A., Patel, J.M.: Efficient aggregation for graph summarization. In: Proceedings of the 2008 ACM-SIGMOD International Conference Management of Data (SIGMOD'08), pp. 567–580, Vancouver, Canada, June 2008

16. Tian, Y., Patel, J.M.: Interactive graph summarization. In: Yu, P.S., Han, J., Faloutsos, C. (eds.) Link Mining: Models, Algorithms, and Applications, pp. 389–409. Springer, New York (2010)

17. Chen, C., Yan, X., Zhu, F., Han, J., Yu, P.S.: Graph OLAP: towards online analytical processing on graphs. In: ICDM, pp. 103–112 (2008)

18. Li, C., Yu, P.S., Zhao, L., Xie, Y., Lin, W.: InfoNetOLAPer: integrating InfoNetWarehouse and InfoNetCube with InfoNetOLAP. PVLDB **4**(12), 1422–1425 (2011)

19. Qu, Q., Zhu, F., Yan, X., Han, J., Yu, P.S., Li, H.: Efficient topological OLAP on information networks. In: DASFAA'11, Hong Kong, pp. 389–403, April 2011

20. Li, C., Zhao, L., Tang, C.J., Chen, Y., et al.: Modeling, design and implementation of graph OLAPing. J. Softw. **22**(2), 258–268 (2011)

21. Zhao, P., Li, X., Xin, D., Han, J.: Graph cube: on warehousing and OLAP multidimensional networks. In: SIGMOD'11, 12–16 June 2011

22. Zhang, N., Tian, Y., Patel, J.M.: Discovery-driven graph summarization. In: 2010 IEEE 26th International Conference on Data Engineering (ICDE). IEEE (2010)

23. Rodrigues, J.F., Triana, J.M., Faloutos, C., Triana Jr., C.: SuperGraph visualization. In: Proceedings of the 8th IEEE International Symposium on Multimedia, pp. 227–234 (2006)

TRS: A New Structure for Shortest Path Query

Qi Wang$^{(\boxtimes)}$, Junting Jin, and Hong Chen

Renmin University of China, Beijing, China
vicky0105.wq@gmail.com, jjt0901@126.com, chong@ruc.edu.cn

Abstract. Shortest Path Query is being applied to more and more professional scopes, such as social networks and bioinformatics. But the exponential growth of data makes it much more challenging since traditional BFS-based algorithms are hard to scale due to the requirement of huge memory.

Different from the traditional algorithm such as Dijkstra algorithm, our method is based on Depth-First-Search, which first constructs the DFS tree with interval-based encoding, and then isolates non-tree edges to generate the TRS structure for the graph. Shortest path queries between arbitrary nodes are performed upon this structure. The final result could be a detail path with exact path cost. This algorithm is quite easy to scale to large graphs, since the TRS algorithm automatically divide the graph into a set of connected components, each of which has a single TRS structure. Our experiments has proved that the algorithm fits large sparse graph quite well in real world.

Keywords: Information network · Large graph · TRS · Shortest path

1 Introduction

Traditional algorithms for SPQ are mainly based on Breadth-First-Search of the graph. The classical Dijkstra's algorithm [8] with O(n^3) time cost and O(n^3) space cost is one of them. And these methods are all memory based, and show great limitations in scalability with the fast growth of data, especially the web data. However, now even a small community network may have thousands of nodes, and storing all of the shortest path trees for this large graph is infeasible at all.

In general, graphs for real large networks are mostly sparse. That is, the average degree of each node is close to 1 or even lower. Accordingly, the storage structure also changes to storing edges instead of storing adjacency matrix. Respect to this property, we proposed a new approach to find the shortest path in sparse graph, and the time cost only associated with the number of edges, while a little more space is needed.

This work is supported by a grant from "Special Research on Key Technology of Domestic Database with High Performance and High Security" for National Core-High-Base Major Project of China (No. 2010ZX01042-001-002-002).

S. Zhou and Z. Wu (Eds.): ADMA 2012 Workshops, CCIS 387, pp. 13–26, 2013.
DOI: 10.1007/978-3-642-41629-3_2, © Springer-Verlag Berlin Heidelberg 2013

Our algorithm first call Depth-First-Search (DFS) algorithm on a directed acyclic graph(DAG), and converts the graph into our new structure TRS(s), in which each node has its interval-based three tuple code. TRS divides all edges into three sets: DFS spanning tree edges (TE), separation edges (SE), and remaining edges (RE). For each TRS, if it only contains DFS spanning tree edges, answering any shortest path query upon it only takes $O(n)$ time if the code is beforehand sorted; If either separation edges or remaining edges are not null, the time cost only depends on the number of non-tree edges. And this is why our algorithm is quite efficient on sparse graphs.

There are three main contributions of our work in total. The first one is we introduce Interval Based Encoding into SPQ to prune some impossible directions; the second one is we proposed a new structure TRS to represent a graph, and TRS makes it possible to answer any SPQ with space cost of $O(n+m)$, where n is number of nodes and m is number of edges; and the last one is our algorithm scales well by using divide-and-conquer strategy. This method is suitable for parallel processing system for big data like hadoop.

The rest of this paper is organized as follows: Sect. 2 depicts related work, then Sect. 3 introduces our TRS framework. In Sect. 4 we give out the full view of TRS algorithm for answering shortest path queries. Section 5 provides the complexity in both time and space for our algorithm. Section 6 shows all the experiments. Section 7 is conclusions and future work.

2   Related Work

The very famous algorithm for shortest path query is Dijkstra [8], and some improvement could be made by using heap data structure for priority queues [9]. Although this algorithm has been extended to external memory [4], it cannot handle well with respect to response time. In order to deal with SPQ, Agrawal and Jagadish [10] introduced the idea of graph partitioning. Later, [7] gave a more efficient way by materializing some local shortest path which could be held in memory. R. Gutman [5] used the reach value of each node to find the shortest path for road networks. Depending on the basic idea of reach value, AV. Goldberg, H. Kaplan and R. Werneck [3] introduced several variants of the reach algorithm, and some do not need explicit lower bounds. Since in the search process, some unrelated branches may be taken into account and it may cause a lot of time cost. Subsequently, methods using geometric attributes of a graph are proposed, and a lot of work has been done. Reference [6] used a geometric way to prune some unrelated branches in order to narrow the search. And [14] divides the graph into several regions, and put an edge into the priority queue in Dijkstra's algorithm when a path from the source region to the destination region passes this edge. The most recent research using geometric attribute is [2], which exploits symmetry of the graphs to compress BFS-trees. Experiments shows that the space cost is reduced in comparison with the un-compressed BFS-trees, but still, space cost is too high when the graph is large enough. Another new research done by F. Wei [1] proposed TEDI as an indexing and query processing scheme for SPQ. Time cost for index construction is $O(n^2)$, while query time depends on the argument of the decomposition.

3   TRS Framework

3.1   Background Knowledge

In this paper, our research object is Directed-Acyclic-Graph with non-negative weight. Let n be the number of nodes, and m be the number of edges. For a complete graph, there's an arc between them for any two nodes, thus number of edges is $M(n) = n * (n - 1)$. But in most graphs, number of edges is much less than $M(n)$, so we define the ratio of m to $M(n)$ as the saturation of G, that is, $saturation(G) = m/n * (n - 1)$. Suppose the saturation of a graph is lower than a given low ratio, e.g. 0.5 %, then we call this graph a sparse graph.

Enlightened by the sparse feature and pruning idea, we propose the TRS structure. In this structure, we classify edges into three sets, TE, SE, and RE. Edges in TE form the spanning tree(or forest) of the original graph. When we only consider the TE edges, if node u could reach node v, the only path is already the shortest path. But at most time, SE and RE must be considered. If the graph is sparse, the number of edges is less than $0.005n^2$. Thus the number of edges in SE and RE is also small, so that the additional time for searching in SE and RE will not be too much. In our experiments, this has been well proved.

3.2   TRS Structure

In this section, we will first introduce Interval-Based Encoding.

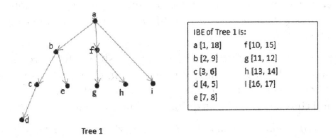

Tree 1

Fig. 1. Example of interval based encoding

Interval-Based Encoding. Interval-Based Encoding was first put forward to be applied to XML files [11], then it is used to solve reachability problems [12,13] well. Interval-Based Encoding utilizes Depth-First-Search to assign both the pre-order code and post-order code for every node of a tree. When a node is visited at the first time it is assigned the pre-order code, and it gets post-order code after all its offsprings are visited. Figure 1 shows an example of Interval-Based Encoding of a spanning tree. Relationship between any two nodes could be represented through their intervals, that is, root a has the most wide interval, which contains all the intervals of a's offsprings, and intervals of b and f has no intersections because they belong to different branches.

In Fig. 1, if we want to find the shortest path from node a to i, we can use the IBE of Tree 1 to directly prune the b-branch and f-branch. Because neither of their intervals satisfy the condition containing i's interval. Considering its $O(1)$ time cost, we think of making it useful in SPQ to avoid a BFS which only depends on the weight of an edge. But as the example indicates that the encoding only fits for tree structure, and if we want to take advantage of it to answer SPQ in an arbitrary graph, we must modify the encoding.

TRS for Graph. For an arbitrary directed graph, it has more edges than its corresponding spanning tree or forest which could be interpreted by the Interval-Based Encoding, thus, in order to answer the shortest path query, these non-tree edges must also be recorded to maintain the complete path information of the original graph.

In our encoding process, we first select a root node and then perform a DFS on the graph from this node to get its spanning tree. And then select next root node as such, as a result, spanning forest comes into being. During this process, the interval for each node is also assigned. If the arc head of the current scanning edge directs a closed node, this edge is a non-tree edge.

Theorem 1. *For a non-tree edge, the start node is ancestor or sibling (we call two nodes siblings if they belong to two different subtrees, without regard to the level of subtrees) of the end node.*

Proof. Nodes in the DFS trees have three relationship, which are ancestor, offspring, and sibling. Assume a non-tree edge $e = \langle u, v \rangle$, v is a closed node in the DFS trees. When the current scanning edge is e, u's post-order is bigger than v's post-order, and thus u cannot be offspring of v, but only can be ancestor or sibling.

Theorem 2. *For a non-tree edge, if the start node is a sibling of the end node, its pre-order is bigger than the post-order of the end node.*

Proof. Edge $e = \langle u, v \rangle$ is a non-tree edge, and u is a sibling of v, and their intervals are respectively $[u_a, u_b]$ and $[v_a, v_b]$. We already know from Theorem 1's proof that $u_b > v_b$. Because different nodes' intervals can't overlap, there may be $u_a < v_a$ or $u_a > v_b$. However, if $u_a < v_a$, then $[v_a, v_b] \subset [u_a, u_b]$, and according to Interval-Based Encoding, u should be v's ancestor, which conflicts with their sibling relationship. Thus the relationship between u_a and v_b can only be $u_a > v_b$.

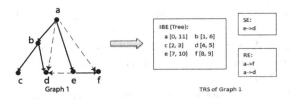

Fig. 2. Example of TRS

According to Theorem 1 and 2, we classify all non-tree edges into two classes: SE(Separation Edges) and RE(Remaining Edges). SE are those edges whose two endpoints are siblings in the DFS tree, and connect two different branches. RE are those edges whose arc tail is ancestor of head in the DFS tree. We store the Interval-Based Encoding instead of the whole DFS tree because the code actually presents the topology of the tree, and the location of each node can be deduced from the codes. So, as to a spanning tree with tree edges and nodes' codes, as well as SE and RE, we call it a TRS structure, reconstructing the graph and keeping all topology information except the weights. Figure 2 is an example.

Weight information is also easy to maintain by adding another item in the Interval-Based Encoding. For each node, we assign it a weight as the sum of weights along the path on the DFS tree from the root node. E.g. in Fig. 2, $w(b) = w(ab)$, in which $w(b)$ represents the weight of node b, and $w(ab)$ represents the weight of edge $\langle a, b \rangle$. If one node are reachable from the other node in a tree, and if we want to get the path weight between the two nodes in the DFS tree,

---

Algorithm 1 TRS Construction

INPUT: DAG G
OUTPUT: TRS(s) for G
ALGORITHM:
01:   Find root nodes set R of G
02:   For 1..n
03:       visited[n]=false
04:   End
05:   For each *root* ∈ R Do
06:       start DFS from root
07:       If v is not visited
08:           visited[v]=true
09:       Else v is boundary node, back to v's parent
10:       End
11:       Assign pre-order if not assigned
12:       Assign post-order if all its offspring is visited
13:   End

Fig. 3. Algorithm of TRS construction

we can directly use the weight of end node subtracting the weight of start node. Because of this feature, we record the path weight from root to the current node as the node weight. And for non-tree edges, we need to record the edge weight. By adding one more item in the code in this way, the weight information of the graph is also maintained.

Sometimes, the graph is not well connected in one spanning tree. For example, if a graph has two connected components, one contains edges $\langle a,b \rangle$ and $\langle a,d \rangle$, and the other contains edge $\langle c,d \rangle$, the DFS scanning would generate a spanning forest rooted at a and c for this graph, and node d is shared by both of them. We call nodes (shared by more than one TRS) boundary nodes like d. In Fig. 3, Algorithm 1 describes how to generate TRS structure for an arbitrary DAG.

4   Answering Shortest Path Query

4.1   The Framework

In general, the DFS process is performed upon a DAG G at first, then it returns the TRS structure of G. If there're more than one spanning tree of G, TRS(G) returns a group of TRS and identified by their root nodes. In each TRS, the nodes as well as their codes, the responding SE and RE are recorded. For each TRS of G, we call AnswerSPQ algorithm to deal with and return the shortest path between the query nodes. When there are more than one TRS structure in a connected DAG, there must be some boundary nodes as intermediate nodes to connect adjoining TRS. For example, If node u is the current encoding node, and it has already been visited and belongs to another TRS, we mark u a boundary node and set it to be closed, so that its offspring (if exist) will not appear in the current TRS, and this avoids mark one node repeatly.

4.2   Answer SPQ On TRS

For the given start node s and end node t, if there is no non-tree edges and t's interval is within s's interval, we search in the encoded nodes from s to t, and the shortest path is a sequence which are ordered by their pre-orders of these nodes, with weight of $w(t)-w(s)$. But mostly, neither SE nor RE is null, and shortest paths take the same chance to pass both of them. Since it's easy to find a path on the DFS tree, the main problem focuses on paths passing these two sets. When two query nodes are reachable, and no SE edges exist in their branch, we only need to replace some tree paths with RE edges if they have smaller weight. When there are SE edges between the branches which contain the two nodes respectively, we need to find the shortest path between the query nodes and end points of SE edges and connect these parts together to get the global shortest path. The algorithm is depicted in Fig. 4.

The first action in Algorithm 2 is to sort edges in SE. According to Theorem 2, we know that edges starts from a laterly-visited node to a formerly-visited node,

---

Algorithm 2 AnswerSPQ

INPUT: TRS of G, query nodes s, t
OUTPUT: Shortest path from s to t if exists
ALGORITHM:
01:    qsort(SE) by pre-order of start node in descending order
02:    If Reachable(s, t) Do
03:        current = RE-SP(s, t)
04:    Else Do
05:        current.weight = INFINITY
06:    End if
07:    For $e \in SE$ Do
08:        If e's start is reachable from s Do
09:            tmp = SE-SP(s, t, e)
10:            If *tmp.weight < current.weight* Do
11:                current = tmp
12:            End If
13:        End If
14:    End For
15:    return current

Fig. 4. Algorithm of AnswerSPQ

and this sorting makes the search starts from the right-most SE edge in a DFS tree (if the tree spreads from left to right) to guarantee no route is missed. If the two nodes are reachable in the DFS tree, we find out a temporary shortest path by taking RE into account, or set them unreachable if not reachable. Then SE is considered. The searching starts from each SE edge whose start node is reachable from s, and when the process gets to the end node of this edge, we treat the end node as a new source node to run SE-SP, until it reaches t or has searched all edges in SE but not reaching t. Algorithm 3 and 4 illustrate the process of RE-SP and SE-SP.

Algorithm 3 only consider RE edges while processing SPQ, because whether or not the shortest path passes SE edges, it needs to consider both DFS tree and RE. Here we offer a schematic figure to explain the relationship of RE edges that appear on the same branch of DFS tree as Fig. 6.

The bold line represents a trunk of DFS, and the arc lines represent RE edges on this branch. Here we must make sure that each RE edge is valuable, that is, for $(a \rightarrow b) \in RE$, $w(a \rightarrow b) < w(b) - w(a)$. Because only when it is valuable, shortest path may pass it, if not, the shortest path chooses the tree path instead of this RE edge. The schema indicates that there are two kinds of relationship between two RE edges: serial and overlying. In Fig. 6, edge $(a \rightarrow b)$ and edge $(e \rightarrow f)$ are serial; edge $(a \rightarrow b)$ and edge$(c \rightarrow d)$, as well as edge $(e \rightarrow f)$ and edge $(g \rightarrow h)$ are overlying. Algorithm 3 describes how to answer shortest path query under the two situations.

Algorithm 3 Shortest Path only Considering RE (RE-SP)

INPUT: query nodes s, t
OUTPUT: Shortest path from s to t
ALGORITHM:
01: Find RE Candidate whose end points are on the path from s to t
      on DFS-Tree
02: qsort(RE Candidate) by pre-order of end nodes in ascending order
03: Initial path queue Q, Q.push(s, s)
04: For each RE $edge(rs, re) \in RECandidates$ Do
05:      Current = path(s, rs)+edge(rs, re)
06:      While !Q.empty Do
07:          Sp(s, t') = Q.pop()
      //Pop Q's first element as the shortest path from s to t in DFS tree.
08:          If (rs, re) overlap sp(s, t') Do
09:              tmp = sp(s, t')+path(t', re)
10:          Else if (rs, re) is serial to sp(s t') Do
11:              $tmp = sp(s, t') + path(t', rs) + path(rs, re)$
12:              brk = true
13:          End If
14:          If $tmp.weight < current.weight$ Do
15:              current = tmp
16:          End If
17:          Q.push(current)
18:          If $brk == true$ break
19:      End While
20: End For
21 sp(s, t')=Q.pop()
22: return $sp(s, t') + path(t', t)$

Fig. 5. Algorithm of RE-SP

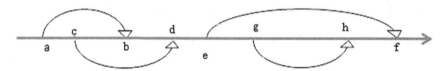

Fig. 6. Schema of RE-SP

Then let's consider the SE edges. From Algorithm 2 we can get a very naive thought, that is, starting from the right most SE edge, and then treating the end of this SE edge as a new query source node, repeating this procedure until we find out the path. But this will lead to traversing all SE edges, so we do a one-hop ahead computation, that is, connecting SE edges who are reachable.

Definition 1. *For any two SE edges, se_1 and se_2, if end of se_1 can reach start node of se_2, we can define se_2 is a SE-child of se_1.*

Algorithm 4 Shortest Path Considering SE (SE-SP)

INPUT: query nodes s, t, SE edge se
OUTPUT: Shortest path from s to t
ALGORITHM:
01: result.weight = INFINITY
02: part1 = RE-SP(s, se.start)
03: part1=$part1 + se$
04: For $p \in$ SE-children Path Array Do
05:     If Reachable(se.end, p.start) Do
06:         pse = first se edge of p
07:         part2 = SE-SP(se.end, t, pse)
08:     End If
09:     If $part2.weight < result.weight$ Do
10:     If Reachable(se.end, t) Do
11:         part3 = RE-SP(part2.last, t)
    //part2.last stands for the end point of the last se edge of p
12:     End If
13:         If $part1.weight + part2.weight + part3.weight$<result.weight Do
14:             result = $part1 + part2 + part3$
15:         End If
16:     End If
17: End For
18: Return result

Fig. 7. Algorithm of SE-SP

With this definition, for an SE edge, we first find out all its SE-children. Then compute the shortest path from its end node to the start nodes of its SE-children by calling algorithm RE-SP. Next, add its own weight to get the shortest path from its start node to the start nodes of its SE-children, and store them in a path array. After all these shortest paths are attained for each SE edge, we order them by the pre-order of start nodes in descending order. When a query comes, we only need to search in this array, until we find out a path that could reach t. Algorithm 4 shows the detail of using this two-hop shortest paths.

If there are more than one TRS structures in a connected DAG, there must be some boundary nodes connecting them. These boundary nodes help to find out the shortest path. If two nodes can be connected through some boundary nodes, we get these paths and select the shortest one as the final result. Some paths may have only one boundary node lies on, while some may have more than one, in this case, shortest path between boundary nodes can be pre-computed.

5   Algorithm Analysis

In this section, we will give the mathematic analysis on the complexity of our algorithm using TRS structures. According to Algorithm 2, time cost for computing shortest path on TRS mainly focuses on SE-SP, let's name it T(SE-SP), and SE-SP calls itself recursively. But we cannot use this recursive algorithm to derive time complexity because for each SE edge, the number of SE edges which it could reach (we call these SE edges the child edge) is unclear, and actually, the use of SE is to enumerate all possible paths. So we utilize the average child SE edges to get the result. Let the average number of child edge a SE edge has be a, and all possible paths to reach the destination node forms an a-tree, and the number of leaf node is the number of searching paths. If the a-tree has k levels, then we get the following equation:

$$1 + a + a^2 + ... + a^{k-1} = |SE| \tag{1}$$

From Eq. (1), we can know that

$$a^{k-1} = (|SE| * (a - 1) + 1)/a \tag{2}$$

and this is equal to the number of leaf node in this tree. So the search upon SE edges is totally conducted $(|SE| * (a - 1) + 1)$ times, and each time k SE edges are considered. Also from Algorithm SE-SP, each time a SE edge is counted in, RE-SP is called, so theoretically, for each such path, RE-SP is called (k-1) times, but many of them are called more than once, and the actual number is

$$1 + a + a^2 + a^3 + ... + a^{k-1} \tag{3}$$

which is $|SE| + 1$.

Here, we still use the average assumption that on each branch that is separated by SE edges, the number of RE edges are almost the same, so that in each branch, the number of RE edge is $|RE|/|SE|$. And time cost for RE-SP is $|RE'|^2$($|RE'|$ is the number of RE edges on the current single branch), so that the total time cost for one a-tree path is $(k - 1) * (|RE|/|SE|)^2$. And then we get that

$$T(SE\text{-}SP) = (|SE|(a - 1) + 1) * (k - 1) * |RE|^2/|SE|^2$$

And time complexity is $O(|RE|^2/|SE|)$. Now, let's turn to Algorithm 2, we can induce that time cost for one TRS is $O(|RE|^2)$. And worst condition for fulfilling a path is $|SE| * n$. Therefore, when a graph only has one TRS, the worst time cost for answering shortest path is $O(|RE|^2 * |SE| * n)$, while the best is $O(|RE|^2 * n)$, and for sparse graph, $|RE|$ and $|SE|$ are both much smaller than n. Our experiment also shows that time cost for sparse graph speeds up.

Space cost for TRS is also quite small. $O(n)$ for node encoding, $O(|SE|)$ for SE edges, and $O(|RE|)$ for RE edges. And clearly, the total space cost does not exceed $O(n+m)$. And this significant space saving will show great advantage in large graphs.

6   Experiments

All experiments in this paper were run on a Intel Core 2 2.66GHz PC with 4G memory. Programming language is C/C++.

Table 1. TRS structure on artificial graph

Node no.	Edge no.	Saturation	Time (ms)	Space (KB)	RE no.	SE no.
100	129	0.013	0	2.05	15	15
200	237	0.006	0	4.29	7	31
300	348	0.004	0	6.91	15	34
400	463	0.003	0	8.99	26	38
500	577	0.002	0	11.0	21	57
600	671	0.002	0	13.5	29	43
700	807	0.002	0	15.7	54	54
800	907	0.001	0	17.9	37	71
900	994	0.001	0	19.8	35	60
1000	1152	0.001	0	23.1	78	75

The experiments first record the time and space cost for constructing TRS for graphs. We generated an artificial graph, whose number of nodes ranging from 100 to 1000, and the saturations are mostly lower than 1 %. In the first experiment, the graph contains one TRS. The first three columns of Table 1 are the number of nodes, number of edges and saturation. And the next two columns list the time cost and space cost for TRS construction.

From this table, we can see the time cost and space cost for construction is quite small for graphs with such scale. And other features such as RE number and SE number are also shown in the table.

To test the correctness and efficiency of our TRS structure and algorithm, we designed a group of experiments. To test the correctness, we run AnswerSPQ on TRS and Dijkstra algorithm using a group of identical queries, and then compare the results. Table 2 indicates that our algorithm can find out the correct shortest path between any two pair of nodes if the path exists.

Table 2. Correctness verification of TRS

Query Points(s, t)	Result using Dijkstra	Result using TRS
(221, 88)	<221, 64, 166, 217, 1, 49, 181, 93, 59, 184, 302>	<221, 64, 166, 217, 1, 49, 181, 93, 59, 184, 302>
(118, 82)	<118, 82>	<118, 82>
(37, 359)	<37, 35, 2, 41, 117, 65, 262, 220, 237, 264, 123, 225, 125, 134, 251, 73, 212, 209, 72, 14, 414, 164, 359>	<37, 35, 2, 41, 117, 65, 262, 220, 237, 264, 123, 225, 125, 134, 251, 73, 212, 209, 72, 14, 414, 164, 359>
(48, 483)	<48, 51, 57, 248, 350, 287, 224, 429, 193, 203, 328, 288, 425, 148, 483>	<48, 51, 57, 248, 350, 287, 224, 429, 193, 203, 328, 288, 425, 148, 483>
(444, 439)	<444, 439>	<444, 439>

Table 3. Result on graph

Node no.	Edge no.	Dijkstra (ms)	TRS (ms)
100	129	0.016	0.109
200	237	0.094	0.047
300	348	0.219	0.093
400	463	0.438	0.110
500	577	0.718	0.172
600	671	0.922	0.110
700	807	1.391	0.157
800	907	1.734	0.250
900	994	2.390	0.141
1000	1152	2.860	0.281

To test the efficiency of TRS, 1000 queries using AnswerSPQ on TRS and Dijkstra algorithm respectively are carried on the graph, and the average time cost is recorded as follows in Table 3.

In order to clearly see our improvement, we give out Fig. 8. The blue line is the cost for Dijkstra algorithm and the red dotted line is the cost for TRS. In the figure, we can see when the graph is small, Dijkstra performs better than TRS. But as the size of graph grows, time cost of Dijkstra increases sharply, while our TRS shows a slow rate of rise, and lead to a much better performance than Dijkstra when graph size is 1000.

In addition, we collected the wikipedia data for our real graph experiments. In this data set, node represents the web page and edge represents the link from one page to another. Total number of nodes for this graph is 1100000, total number of edges is 1133321, and the saturation is 0.0000113 %. Construction of this graph totally outputs 37048 TRSs, and 151639 boundary nodes. The biggest TRS contains 5667 nodes and 6547 edges, and 1934 boundary nodes. Table 4 gives the result performing AnswerSPQ on this real graph.

Because the real graph is so large, that it's necessary to make preparation for computing. The pre-computation phase can be divided to three parts. The first is to parse the graph to forests, the second is to generate TRS for each tree in the forest, and the third is to compute local shortest paths between boundary nodes. Since we do not directly store all these information in memory, all of the three parts include disk I/O time cost. The query time 17.5 milliseconds is also the average query time for 1000 different queries, and for such a big graph, the response time is acceptable while the BFS method cost 32.47 milliseconds in a graph contains 592,983 nodes.

Table 4. Result on Wikipedia graph

Node No.	Edge No.	Space (MB)	Pre-computation time (min)	Query time (ms)
1,100,000	1,133,321	36.853	107	17.5

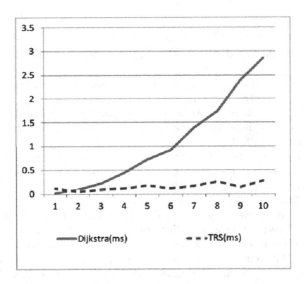

Fig. 8. Result on graph

7   Conclusion and Future Work

Answering shortest path query on a graph which does not fit in memory is now a very important problem with the development of networks. In this paper, we first introduce Interval-Based Encoding into the graph re-construction stage and give out a new structure TRS of the graph. For most graphs, the construction time is short. Proved by experiments, TRS structure performs quite well, and time cost increase almost linear respect to the graph scale. When AnswerSPQ on TRS algorithm computes shortest paths in graph with more than one TRS, this method does not perform quite well when graph is small, but our experiment results shows that, it performs well when graph is large.

In our experiments, we notice that the pre-computation time cost is high, and it mainly consumes on disk I/O and computing local shortest path. For disk I/O, our next attempt is to maintain graph in memory when the scale is not large. And if a TRS contains too many boundary nodes, the cost is high. E.g., in the TRS we mentioned above which contains 1,935 boundary nodes, total 3,744,225 queries need to be answered. So we will focus on how to improve efficiency on graph with a lot of boundary nodes. Besides, the graph tends to be dynamic in real world, how to maintain the query efficiency using TRS structure when either some nodes or edges change remains to be studied further.

References

1. Wei, F.: TEDI: Efficient shortest path query answering on graphs. In: SIGMOD'10 (2010)
2. Xiao, Y., Wu, W., Pei, J., Wang, W., He, Z.: Efficiently indexing shortest paths by exploiting symmetry in graphs. In: EDBT'09 (2009)
3. Goldberg, A.V., Kaplan, H., Werneck, R.: Reach for A*: efficient point-to-point shortest path algorithms. In: Workshop on Algorithm Engineering and Experiments, pp. 129–143 (2006)
4. Arge, L., Meyer, U., Toma, L.: External memory algorithms for diameter and all-pairs shortest-paths on sparse graphs. In: Díaz, J., Karhumäki, J., Lepistö, A., Sannella, D. (eds.) ICALP 2004. LNCS, vol. 3142, pp. 146–157. Springer, Heidelberg (2004)
5. Gutman, R.: Reach-based routing: a new approach to shortest path algorithms optimized for road networks. In: Proceedings of the 6th International Workshop on Algorithm Engineering and Experiments, pp. 100–111 (2004)
6. Wagner, D., Willhalm, T.: Geometric speed-up techniques for finding shortest paths in large sparse graphs. Konstanzer Schriften in Mathematik und Informatik, pp. 1430–3558 (2003)
7. Chan, E.P.F., Zhang, N.: Finding shortest paths in large network systems. Department of Computer Science, University of Waterloo (2001)
8. Dijkstra, E.W.: A note on two problems in connexion with graphs. Numerische Mathematik (1959)
9. Cormen, T.H., Leiserson, C., Rivest, R., Stein, C.: Introduction to Algorithms. MIT Press, Cambridge (2001)
10. Agrawal, R., Jagadish, H.V.: Algorithms for searching massive graphs. IEEE Trans. Knowl. Data Eng. **6**, 225–238 (1994)
11. Eietz, P., Sleator, D.: Two algorithms for maintaining order in a list. In: Proceeding of the19th Annual ACM Symposium on Theory of Computing (STOC), pp. 365–372 (1987)
12. TRßl, S., Leser, U.: Fast and practical indexing and querying of very large graphs. In: SIGMOD'07 (2007)
13. Wang, H., He, H., Yang, J., Yu, P.S., Yu, J. X.: Dual labeling: answering graph reachability queries in constant time. In: Proceedings of the 22nd International Conference on Data Engineering (ICDE), p. 75 (2006)
14. Lauther, U.: An extremely fast: exact algorithm for finding shortest paths in static networks with geographical background. IfGlprints 22, Institut fuer Geoinformatik, Universitaet Muenster (ISBN 3-936616-22-1), pp. 219–230 (2004)

Parallel Simrank Computing on Large Scale Dataset on Mapreduce

Lina Li[✉], Cuiping Li, and Hong Chen

Renmin University of China, Beijing, China
lilina.8888@yahoo.com.cn

Abstract. Many fields need computing the similarity between objects, such as recommendation system, search engine etc. Simrank is one of the simple and intuitive algorithms. It is rigidly based on the random walk theorem. There are three existing iterative ways to compute simrank, however, all of them have one problem, that is time consuming; moreover, with the rapidly growing data on the Internet, we need a novel parallel method to compute simrank on large scale dataset. Hadoop is one of the popular distributed platforms. This paper combines the features of the Hadoop and computes the simrank parallel with different methods, and compars them in the performance.

Keywords: Simrank · Parallel · Mapreduce · Hadoop

1 Background

There are two kinds of methods computing the similarity between two objects, one is based on the content of text. This method takes the text object as a set of short terms or phrases, and the weights of each terms constitute a feature vector. Then use proper method to compute the similarity between the text objects with the feature vector. The other method is computing similarity between linked objects. As the objects are linked, more information can be inferred from the relation between the objects. Simrank is one of the methods to compute similarity between linked objects. And simrank is applied in many fields such as Collaborative Filtering Recommendation, cluster algorithm, classification algorithm and search engines.

1.1 Related Knowledge

Simrank is an important method to compute similarity between objects. This method can be applied to compute the similarity between hyperlink documents, and then sort the documents by the similarity value. This method can also be

This work is supported by National Core-High-Base Major Special Sub-ject'Research on Key technology on High Performance High security Domestic database'(2010ZX01042-001-002).

S. Zhou and Z. Wu (Eds.): ADMA 2012 Workshops, CCIS 387, pp. 27–40, 2013.
DOI: 10.1007/978-3-642-41629-3_3, © Springer-Verlag Berlin Heidelberg 2013

used in recommendation system to recommend an item to a client who might be the potential costumer.

So far, there are three ways to compute simrank:

The first one is a naive formula which is raised by the simrank proposer. This method calculates similarity accurately, but the drawback of this method is the high time complexity: $O(n^3)$, during each iteration step, the space complexity is $O(n^2)$; the second one is matrix multiplication, this method turns the initial formula to a standard adjacent matrix multiplication. The last one is random walk based on the Monte Carlo Theorem, the similarity between two objects is determined by how long it takes two nodes to meet each other in the directed graph. Right now, the existing methods are not scalable and not practical on large dataset.

In the simrank algorithm, it usually abstracts the objects and connections between objects into a directed graph. The node in the graph represents the object, and the directed edge in the graph represents the connection between objects.

Mapreduce is a parallel programming model and Hadoop is one of the implementation. This paper pays more attention to the details of the simrank parallellization on Hadoop platform.

1.1.1   Initial Algorithm Introduction

This method is based on the following suppose: If two objects are similar, then the neighbors of the objects are also similar. The proposer of simrank algorithm Glen Jef gives an initial formula as following:

$$S(a,b) = \frac{c}{|I(a)||I(b)|} \sum_{i=1}^{|I(a)|} \sum_{j=1}^{|I(b)|} S(I_i(a), I_j(b)) \tag{1}$$

In this formula, I(a) is a set of the nodes which are input node of a, $I_j(a)$ is node j in the node set I(a). The same to $I_i(a)$. This formula indicates that the similarity of nodes are mainly determined by their neighbors. At first, the similarity between a node and itself is 1, and the similarity between two different nodes is 0. Then compute iteratively, the iterative formula as following:

$$S^k(a,b) = \frac{c}{|I(a)||I(b)|} \sum_{i=1}^{|I(a)|} \sum_{j=1}^{|I(b)|} S^{k-1}(I_i(a), I_j(b)) \tag{2}$$

1.1.2   Matrix Multiplication Introduction

In paper [11], there is a novel method computing similarity between objects. In this method, the initial formula is turned to be the following format:

$$S^k = c * W^T S^{k-1} W + (1-c) * I \tag{3}$$

In this formula, I is an identity matrix, W is a column standardized matrix of the adjacency matrix of the graph. W^T is the transport matrix of matrix

W. In this method, computing similarity of objects changes to compute matrix multiplication. And the related proof is presented in the paper.

1.1.3 Monte Carlo Method Introduction

Monte Carlo is an import numerical calculation method based on statistical theory, and random number is often used to solve problem.

Firstly, abstract objects and connections of objects to directed graph, object is the node in graph and the connection is the edge in the graph. This method is based on the random walk theorem, the node in graph walks randomly in the reverse graph. If two nodes meets during a random walk process after same walk step, then the two nodes are similar and the similarity is C^k. k is the step the two nodes meet, and C is the delay factor. During one random walk process, two nodes may meet several times.In this case, the similarity is determined by the step the nodes meet at first time.

As this method is based on sampling, there is deviation in the result. In order to improve the accuracy of this algorithm. It is a good idea to increase the number of sampling. Finally, the similarity is the mean value of several sampling processes.

1.1.4 MapReduce Computing Model

MapReduce is a parallel computing model. It splits the input dataset into M parts and starts M mappers, and each mapper deals with one part. Mapper produces key/value pairs and writes them into the intermedia files. If there are N key values in the intermedia result and then start N reducers, finally produce the final results. This is called a MapReduce job.

Designing a MapReduce algorithm has some limitations, such as any computing process should be turned into map/reduce procedures and obey strict rules. The complicity of the algorithm makes it more challenging. So far there are some modified frameworks such as PigLatin, SCOPE. They provide a higher level language and even other operators such as join operator; frameworks like HaLoop, iMapReduce, they are all modified on the initial Hadoop. However in this paper, we dont make any change on Hadoop platform, but we take the strategy iMapReduce uses which will be mentioned in the following section. All implementations are on the original Hadoop platform.

1.1.5 Contribution of this Paper

This paper combines the features of MapReduce Model to compute simrank on the distributed platform. The following sections give detail description about the algorithm and run the experiments on different scale datasets, then analyse the accuracy and efficiency of the results.

2 Simrank Parallel Computing

This section gives a detail description of the algorithm on Hadoop platform with MapReduce Computing Model. When computing Simrank, we abstract an object to a node on the graph and assign each node a unique identification number, the relations between nodes are abstracted directed edges on the graph. Then the objects we will analyse turn to be a directed graph. Directed graph G(V,E), V({v1,v2,v3vn}) is the collection of all nodes in the graph; E({e1,e2,e3em}) is the collection of all edges in the graph.

In the experiment, assign a unique number to all nodes on the graph (start from 1). The node pair (i,j) represents the directed edge, and also indicates that there is a directed edge from node u_i to u_j. O(u) is the collection of all output nodes of node u, I(u) is the collection of all input nodes of node u. In all algorithms mentioned below, record (u,v,sim) means that the similarity of object u and object v is $sim(u < v)$.

2.1   Parallelism of Initial Algorithm

The initial formula of simrank is

$$S^k(a,b) = \frac{c}{|I(a)||I(b)|} \sum_{i=1}^{|I(a)|} \sum_{j=1}^{|I(j))|} S^{k-1}(I_i(a), I_j(b))(k > 0) \qquad (4)$$

Rewrite the formula to the one below:

$$S^k(a,b) = \sum_{i=1}^{|I(a)|} \sum_{j=1}^{|I(b)|} \frac{c}{|I(a)||I(b)|} S^{k-1}(I_i(a), I_j(b))(k > 0) \qquad (5)$$

same to:

$$S^k(a,b) = \sum_{i=1}^{|I(a)|} \sum_{j=1}^{|I(j))|} \frac{c}{|I(a)||I(b)|} S^{k-1}(i,j)(k > 0) \qquad (6)$$

If $i \epsilon I(a)$, $j \epsilon I(b)$, and replace $\frac{c}{|I(a)||I(b)|} S^{k-1}(i,j)$ with $S_{ij}(a,b)$:

$$S_{ij}(a,b) = \frac{c}{|I(a)||I(b))|} S^{k-1}(i,j) \qquad (7)$$

Then the kth iteration of simrank is:

$$S^k(a,b) = \sum_{i=1}^{|I(a)|} \sum_{j=1}^{|I(b))|} S_{ij}^{k-1}(a,b)(i \epsilon I(a), j \epsilon I(b)) \qquad (8)$$

$S_{ij}^{k-1}(a,b)$, represents the similarity contribution of node pair (i,j) making to node pair(a,b) in the kth iteration, and the value is $\frac{c}{|I(a)||I(b))|} S^{k-1}(i,j)$.S(a,b), the similarity of node pair (a,b) is the sum of all contribution made by other node pairs during one iteration.

During one iteration computation, compute all contribution the node pair (i,j) makes to other node pair (a,b) as $S_{ij}(a,b)$. This process can be achieved by a parallel map procedure, and the key of the map output is $(a,b)(a < b)$, the output record is like $(< a,b >, S_{ij}(a,b))$. With the features of MapReduce Computing Model, the records with a same key will be pushed to a same reducer, the reduce procedure sums up all the contribution of one same pair. In the kth iteration step, the similarity of last iteration $S^{k-1}(i,j)$ of any node pair is known, then the similarity of $S_{ij}(a,b)$ is easy to be computed. One iteration step uses one MapReduce job.

In the computing process, each computer in the distribute platform should obtain the topology of the whole graph, the input and output node collections of a node, this information is recorded as (u in:a,b,c—out:x,y,z), which is written in a file and sent to each computer. As the graph is static, the information of the whole graph is static too. In the implementation, caching the graph information in the Distributed Cache, this can reduce the communicate consumption between tasktracker nodes and the jobtracker node. When running a MapReduce job for the first time, cache the topology information in the memory and keep it during the computing process which can be used in all reduce procedure, in this way, it can reduce the expenses of swap, and reduce the communication time through network.

At first, preprocess the topology of the graph, change the initial record (u,v) which represents the directed edge in graph to the format $(< u >, in : a, b, c | out : x, y, z.)$, then gets the inputs and outputs of all nodes. The preprocessing in MapReduce Computing Model is described as following: Map procedure: turn the input record (u,v) to the records $(< u >, out : v)$ and $(< v >, in : u)$. Reduce procedure: reduce the records with a same key to a record as $(< u > , in : a, b, c | out : x, y, z)$.

After the pre-process, distribute the final result files to all computers in the platform, which will be used in the following iterative computing. The input file of each iteration step is the output of the last iteration.

In the beginning, the similarity between a node and itself is 1, so initially the similarity record is like (u,u,1). In order to compute parallel on Hadoop platform, there is an optimize strategy, spliting the input file into several small files then each map processing one small file.

The description of the iteration process is as below:

Map procedure: In the preprocess, the topology of the graph has been stored in every computer. The input record is (u,v,sim), with records $(< v >, out : * | in : *)$ and $< v >, out : * | in : *)$ cached in the memory, and can figure out the contribution value $S_{uv}(i,j)$ the node pair (u,v) makes to other node pairs (i,j). The result is recorded as format $(< i, j >, sim)$.

Reduce procedure: Summing up records $(< u, v >, sim)$ within the same key, then the similarity of node pair (u,v) in this iteration is figured out.

This is the description of a whole iterate process, each iteration starts a MapReduce job.

2.2   Parallel Matrix Multiplication Method

The core of this method is matrix multiplication. Let us analyse the matrix multiplication first.

2.2.1   Single Tuple Method (STM)

Matrix A multiplies Matrix B, and the result Matrix is C: $c_{ij} = \sum_{t=1}^{m} a_{it}b_{tj}$

Suppose that both Matrix A and Matrix B have m columns and m rows, the value of element in i^{th} row and j^{th} column is t, and it is represented as a record (i,j,t)(t not equals 0).

In the pre-process, each element in Matrix A is represented as record $(j, 1, i, j, a_{ij})$, each element in Matrix B is represented as record$(i, 2, i, j, b_{ij})$. The description of pre-process is as below:

Map procedure: turn the record (i, j, a_{ij}) of Matrix A to $(< j >, 1, i, j, a_{ij})$; turn the record (i, j, b_{ij}) of Matrix B to $(< i >, 2, i, j, b_{ij})$.

Reduce procedure:no reduce.

In this process, records are divided into two kinds $(< j >, 1, i, j, a_{ij})$ and $(< i >, 2, i, j, b_{ij})$. The records with same key are sent to a same reduce. In a reduce procedure, the element record $(< t >, 1, m, t, a_{mt})$ from matrix A as a_{mt} and element record $(< t >, 2, t, n, b_{tn})$ from matrix B as b_{tn}, multiply them $c_{mn} = a_{mt} * b_{tn}$, c_{mn} is an element of the result matrix, recorded as $(< m, n >, c_{mn})$. Then start another MapReduce job to sum up all c_{mn} with the same key. Then pre-process the result Matrix c, which will be used as input of the next iteration.

Matrix Multiplication needs two MapReduce jobs. One job multiply each element from Matrix A by an element from Matrix B and the other job is to sum them up.

Apply the method mentioned to computing the simrank, the formula of Simrank is: $S^k = c * W^T S^{k-1} W + (1-c) * I$. W is the column standardized adjacency matrix, c is delay factor.

When computing the simrank formula, take the $(c * W^T) as W_c^T$, $(1 - c) * I$ as I_{1-c}, and then the formula turns to $S^k = W_c T S^{k-1} W + I_{1-c}$

For example, suppose adjacency matrix is:

$$\begin{pmatrix} 0 & 1 & 1 & 0 & 0 \\ 0 & 0 & 1 & 1 & 1 \\ 1 & 0 & 0 & 1 & 0 \\ 1 & 1 & 1 & 0 & 0 \\ 1 & 0 & 1 & 0 & 0 \end{pmatrix}$$

After Column Standardizing, the matrix is:

$$\begin{pmatrix} 0 & 0.5 & 0.25 & 0 & 0 \\ 0 & 0 & 0.25 & 0.5 & 1.0 \\ 0.33 & 0 & 0 & 0.5 & 0 \\ 0.33 & 0.5 & 0.25 & 0 & 0 \\ 0.33 & 0 & 0.25 & 0 & 0 \end{pmatrix}$$

Applying the Matrix multiplication method on MapReduce to compute simrank, Matrix W is represented as a collection of records like (1,2,0.5), and matrix W_c^T is represented as a collection of records like (2,1,0.5*c).

As the definition of Simrank, the similarity between a node and itself is 1, so S^0 is:

$$\begin{pmatrix} 1 & 0 & 0 & 0 & 0 \\ 0 & 1 & 0 & 0 & 0 \\ 0 & 0 & 1 & 0 & 0 \\ 0 & 0 & 0 & 1 & 0 \\ 0 & 0 & 0 & 0 & 1 \end{pmatrix}$$

Matrix I is represented a collection of records like (1,1,1), Matrix I_{1-c} is a collection of records like $(1,1,1-c)$.

At first, mulptily Matrix W_c^T by Matrix S, the result is Matrix T, then multiply Matrix T by Matrix W and add result matrix to Matrix I_{1-c}. At this time, one iteration step finish. As the graph is static, matrix W_c^T and matrix W in the formula $S^k = W_c^T S^{k-1} W + I_{1-c}$ is unchanged during the process. In preprocess, get Matrix W_c^T, W and I ready.

2.2.2   Row Divided Method (RDM)

There is another method of matrix multiplication, which is more effective than the one before. For example: multiply matrix A by matrix B, as below:

$$\begin{pmatrix} A_{11} & A_{12}... & A_{1n} \\ A_{21} & A_{22}... & A_{2n} \\ ... \\ A_{m1} & A_{m2}... & A_{mn} \end{pmatrix} \times \begin{pmatrix} B_{11} & B_{12}... & B_{1t} \\ B_{21} & B_{22}... & B_{2t} \\ ... \\ B_{n1} & B_{n2}... & B_{nt} \end{pmatrix}$$

From the matrix multiplication formula, the element c_{ij} in the result Matrix C is vector multiplication result of ith row in Matrix A and j^{th} column in Matrix B. Each row in Matrix A multiply by different column in B is independent, which can be parallel.

And here we do not divide Matrix B, each computer in distribute platform keeps the records of Matrix B in memory. The process of matrix multiplication is as following:

Map procedure: turn input record (i, j, a_{ij}) to record $(<i>, j, a_{ij})$.

Reduce procedure: from input records $(<i>, j, a_{ij})$, get the row record in Matrix A as record $(<i>, 1, a_{i1}; 2, a_{i2}; 3, a_{i3}...)$.Multiply the row record to each column of the Matrix B.

Several experiments show this method is more effective than the single tuple method.

The division of Row Divided Method is based on row unit of a matrix and the division of Single Tuple Method is based on unit of tuple of a matrix. Row Divided Method produces less intermedia records than the Single Tuple Method, and it reduces the expenses of sort in shuffle procedure.

In the Single Tuple Method, there are (m*n+n*t) tuples in matrix A and matrix B, so there are (m*n+n*t) records after map procedure. In the shuffle procedure, all these records are sorted and sent to different reducers. In the Row Divided Method, it only divides the matrix A by row and doesnt divide matrix B. In the shuffle procedure, it sorts only m records, which is much less than the sorted records in Single Tuple Method. It reduces the expenses during sorting and web communication.

2.2.3   Comparison Between Two Methods

The difference between the two methods mentioned before is the division granularity. Single Tuple Method is divided by element unit, and Row Divided Method is divided by row unit. As we expected, the first method should be more effective than the second one, because it is more parallel than the second one. But the result of the experiments is totally opposite, which shocks us.

Then we analyse the amazing results: In the parallel MapReduce Computing Model, there are data exchanging procedures between map procedure and reduce procedure, which are sort and shuffle procedures. The sort procedure is sorting all the outputs of map procedure, and it needs communication between different nodes. When finishing sorting, shuffle procedure sends the sorted results to the different reduce. The time consuming in sorting cant be ignored. High level parallel is important but the consuming of the sort and shuffle procedures cant be neglected. The efficiency is determined by several factors, high level parallelism is only one among them.

2.3   Parallel Monte Carlo Method

Monte Carlo Method is based on the random walk theorem. In this method, each node in the directed graph walks to its neighbours randomly. After several random walk steps, each node has its own walk track. In all the tracks, there are nodes meeting after a same step length. Suppose node u and node v meets after t steps, the similarity between the nodes is c^t. This process is called one random walk sampling. In order to guarantee the accuracy, it is necessary to sample more times and take the average of the results as the similarity of nodes.

Start from node u in the graph, pick one node t in the O(u), and form the 1-step random walk track (u,t) and samples n times like this. So there are n 1-step random walk tracks like (u,v) from node u, so does the node v like (v,w). Join the 1-step walk tracks, we can infer the 2-step walk track (u,w). After t iterations, the results are 2t-steps tracks of all nodes. From all the random walk tracks, in the tth sampling, if node u walks to node s recorded as (i,t,s,u). Then it is easy to figure out the meetings of node pair during random walk. To compute the similarity of node pair should find out step length the first time they meet. The step lengths they meet first time determine the similarity of the node pair.

The Monte Carlo Method has two steps:

1. Generate the random walk tracks
2. Figure out step length node pair meet the first time

From all the directed edges, we get a collection of output nodes of node u, O(u). This is the same as the description before.

Now, we introduce the method of generating random walk track of each node using MapReduce Computing Model, str(x) is a walk track which begin at node x and end at node y.

Map procedure: the input record is (x,y,str(x)), and output records are $(< x >, [x, y, str(x)])$ and $(< y >, [x, y, str(x)])$;

Reduce procedure: join two tracks together, for example: the input records are $(< y >, x, y, str(x))$ and $(< y >, y, z, str(y))$, the output record is $(< x, z >, str(x))$.

In this random walk method, after an iteration, a k-step track turns to a 2k-step track. Compared with original random walk method which walks one step each iteration, this method reduces the iteration times when walking a same step length. After generating all random walk tracks, analyse all the tracks and figure out the steps node pair meet first time. The process is as following:

Map procedure: the input record is (a,b,str(a)) which represents the random walk of node a. It generates the record $(< n, t, a >, s)$. It means that at t^{th} sampling time and at nth step, node n arrives at node s.

Reduce procedure: The input records are like $(< n, t, a >, s)$. At the tth sampling, node s_i and node s_j meet at step n, the output record is (s_i, s_j, t, n).

During one sampling time, node pair may meet more than once, and the step length of the first time they meet is useful to the similarity of the node pair. The method to find the first time a node pair meeting is as following:

Map procedure: input record is (s_i, s_j, t, n), and output record is like $(< s_i, s_j, t >, n)$, point $< s_i, s_j, t >$ as the key of the record.

Reduce procedure: input record is $(< s_i, s_j, t >, n)$, its easy to find the minimum n among the records with the same key. The minimum n is the step length a node pair meets first time.

After figuring out the step length node pairs meet first time in one sampling process, so does the similarity of the node pair in one sampling. In order to improve the accuracy, more samples should be taken.

3   The Result of Experiments

This paper realizes all the algorithms of computing simrank on a Hadoop platform with 20 slave machines and a master machine. Each machine has 6 cores, and installs Linux of Redhat Enterprise version. All the experiments are developed by JAVA language and using hadoop API. Four methods are tested on five different scale datasets, and comparing the efficiency of the different methods. The dataset comes from Stanford.

3.1   Initial Formula Method

As the MapReduce splits the input files of map procedure into small files by 64M and each split is consumed by a map procedure. In the experiment, the input file of map procedure is less than 64M, so only one map will work and it is not parallel. In order to start more map procedures, we split the input file into several small files which is far samller than 64M, and then many map procedures will work parallel. Comparing the parallel Initial Formula simrank method with existing initial methods, the new method is more effective and scalable.

3.2 Matrix Multiplication Method

3.3 Monte Carlo Method

3.3.1 Accuracy Analysis of the Monte Carlo Method

As the method is based on sampling, there is deviation. The walking step length and sampling time influence the accuracy of the results. If walking the same step length, the more sample time is the more accurate the result is. However, if sampling time keeps the same, the longer the step length is the more accurate the result is, but when the length exceeds a limitation, the accuracy doesnt improve any more. The limitation step length is determined by the diameter of the graph.

When using this method computing simrank, it is important to choose a proper step length. If the length is too short, a random walk track will not cover some nodes, and if the length is too long there are circles in the walking track.

The following experiment is supposed to confirm the influence the sampling time to the accuracy. As comparing the accuracy on a big graph is a difficult job, a small graph is chosen here. There are only five nodes on the graph, see as bellow:

First, convert the original graph to a reverse graph:

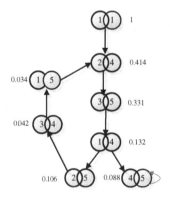

Then begin random walk on the reverse graph, the walking length is 8, and 3 iterations is enough. One experiment samples 10 times and the other does 20 times. Each case is tested 10 times and figures out the deviation. The similarity calculated by Monte Carlo method is marked as S_{random}, and the standard similarity is S_{init}, the deviation is defined as:

$$r = \frac{S_{random} - S_{init}}{S_{init}} \tag{9}$$

r is used to judge the deviation, the table below shows under two cases, in each experiment the number of node pair whose r is larger than 0.5.

Table 1. The number of node pairs whose r is larger than 0.5

Experiment run count	1	2	3	4	5	6	7	8	9	10
Sampling 10 times	4	6	5	2	4	2	5	2	4	3
Sampling 20 times	4	3	4	2	2	4	3	5	3	2

Table 2. Number of the experiments

The number of node pair	2	3	4	5	6
Sampling 10 times	3	1	3	2	1
Sampling 20 times	3	3	3	1	0

Each case runs 10 times, the following table shows the number of the experiments in which the number of node pair which r is larger than 0.5 is a definite value.

In this experiment, we find that: with the times of the sampling increases, the accuracy is improving.

3.4 Comparison Between the Several Methods

The four different algorithms of Simrank is tested on six different scale datasets, and compare the difference between them. Table 8 shows the features of each datasets. Wiki-vote dataset contains 7115 nodes and 103689 edges, details showed in Table 3. P2P-Gnutella05 dataset contains 8846 nodes and 31839 edges, details showed in Table 4. P2P-Gnutella31 dataset contains 62586 nodes and 147894 edges, details showed in Table 5. Web-Stanford dataset contains 281973 nodes and 2312497 edges, as the graph is big that less iteration is carried out. Wiki-talk dataset is too large that only RDM can give the final results, details showed in Table 7.

The tables above show the result of the four experiments on different scale datasets. Monte Carlo Method is the most effective among them but the accuracy

Table 3. Wiki-vote dataset

	Pre-process	1st iteration	2nd iteration	3rd iteration	4th iteration	Final analysis
Initial	1 min30 s	3 min26 s	1 h39 min4 s	1 h57 min42 s	2 h2 min14 s	-
STM	4 min12 s	6 min33 s	7 min29 s	8 min26 s	8 min32 s	-
RDM	1 min6 s	2 min13 s	2 min13 s	2 min16 s	2 min18 s	-
Monte Carlo	1 min1 s	1 min2 s	1 min2 s	1 min1 s	1 min2 s	4 min57 s

Table 4. P2P-Gnutella05 dataset

	Pre-process	1st iteration	2nd iteration	3rd iteration	4th iteration	Final analysis
Initial	1 min30 s	1 min6 s	1 min8 s	2 min15 s	8 min14 s	-
STM .	6 min14 s	9 min38 s	9 min34 s	9 min36 s	10 min8 s	-
RDM	1 min6 s	2 min12 s	2 min12 s	2 min15 s	2 min12 s	-
Monte Carlo	1 min33 s	1 min34 s	1 min35 s	1 min34 s	1 min33 s	6 min4 s

Table 5. P2P-Gnutella31 dataset

	Pre-process	1st iteration	2nd iteration	3rd iteration	Final analysis
Initial	1 min33 s	1 min36 s	7 min28 s	1 h23 min11 s	-
STM	6 min56 s	9 min28 s	12 min23 s	12 min	-
RDM	1 min10 s	2 min45 s	2 min43 s	3 min2 s	-
Monte Carlo	1 min3 s	1 min2 s	1 min2 s	1 min1 s	6 min37 s

Table 6. Web-Stanford dataset

	Pre-process	1st iteration	2nd iteration	3rd iteration	Final Analysis
Initial	1 min57 s	2 h2 min37 s	104 h49 min58 s	-	-
STM	3 min20 s	7 min1 s	88 min32 s	5 h42 min34 s	-
RDM	1 min18 s	29 min55 s	23 min50 s	27 min25 s	-
Monte Carlo	2 min22 s	2 min20 s	2 min19 s	2 min20 s	17 min40 s

Table 7. Wiki-talk dataset

	Pre-process	1st iteration	2nd iteration	3rd iteration	4rd iteration	5th iteration	6th iteration
RDM	1 min 28 s	6 h38 min 28 s	5 h47 min 54 s	5 h45 min 6 s	5 h31 min 3 s	5 h29 min 12 s	5 h31 min 51 s

Table 8. Features of all datasets

	Nodes	Edges	Diameter	Average degree
Wiki-vote	7115	103689	7	14.6
P2P-Gnutella05	8846	31839	9	3.6
P2P-Gnutella31	62586	147892	11	2.4
Web-Stanford	281903	2,312,497	740	8.2
Wiki-Talk	2,394,385	5,021,410	9	4

of this method isnt the same as others. But increasing the time of sampling, the accurate is improving.

The efficiency of different methods are determined by several factors, such as the number of nodes the number of edges average degree of each node in the graph and so on. For example, the number of nodes in P2P-Gnutella05 is larger than the number of nodes in wiki-vote, but the edges in P2P-Gnutella05 is less than the edges in wiki-vote, dataset. In the Initial Formula method, the P2P-Gnutella01 is faster than the dataset of wiki-vote.

There are millions of nodes in the wiki-talk dataset, except the RDM can deal with this kind scale of dataset, others are not suitable for the dataset of this scale. RDM has a better scalability. In the existing method, no one can deal with this kind of dataset.

4   Related Work

MapReduce Computing Model is widely used in several application, such as machine learning data mining text processing and so on. MapReduce is a good tool for processing large scale graph. This paper designs several new algorithms for simrank on MapReduce, including complicate Initial Formula Method Monte Carlo Method and Matrix Multiplication. Combine with the features of the MapReduce Computing Model, reduce the time consuming and improve the parallelism. After several experiments, the Row Divided Method is an excellent method for large scale dataset.

In the paper [9], author gave the definition of simrank and the formula of simrank. After that several novel methods were brought out. Fegaras and Racz [7] proposed a new algorithm on computing simrank which is Monte Carlo Method based on random walk thereom, this is totally different from the formers. In [8], there is another method, in this method, it caches the partial sum of some similarity which will be used often later. This reduces the time consuming for the replicate computing. References [11,12] use matrix multiplication method to compute the simrank. On the other hand, there are some new platform based on the distributed platform Hadoop, such as iMapReduce HaLoop.

5   Conclusion

This paper combines with the feature of MapReduce Computing Model ,designs several novel methods for simrank computing, which not only improve the efficiency but also can deal with large scale dataset. By comparing the several different methods, we find that Monte Carlo Method is the fastest among all the parallel methods, Initial Formula Method is not that quick but it ensures the accuracy. Compared with the existing initial method, this parallel method is improved much in efficiency. Matrix multiplication method has good scalability especially the Row Divide method. This strategy can also be applied to other matrix algorithm.

6 Feature Work   ·

Optimize the parallel methods mentioned in this paper, and adjust several kinds of parameters of the Hadoop platform to make best use of it to accelerate the parallel computing. Or to modify the Hadoop platform to a new one which is suitable for the simrank computing that is another solution. Besides making some changes on the distributed platform, split the graph to small graphs and compute parallel is another solution.

References

1. Dean, J., Ghemawat, S.: MapReduce: simplified data processing on large clusters. In: OSDI 04: Proceedings of the 6th Conference on Symposium on Opearting Systems Design and Implementation (2004)
2. Zheng, Y., Gao, Q., Gao, L., Wang, C.: iMapReduce: a distributed computing framework for iterative computation
3. Bu, Y., Howe, B., Balazinska, M., Ernst, M.D.: HaLoop: efficient iterative data processing on large clusters. Proc. VLDB Endownment **3**(1), 285–296 (2010)
4. Kambatla, K., Rapolu, N., Jagannathan, S., Grama, A.: Asynchronous algorithm in MapReduce. In: 2010 IEEE International Conference on Cluster Computing
5. Cohen, J.: Graph twiddling in a MapReduce world. Comput. Sci. Eng. **11**(4), 29–41 (2009)
6. Bahmani, B., Chakrabarti, K., Xin, D.: Fast personalized PageRank on MapReduce. In: SIGMOD 11, 12–16 June 2011, Athens, Greece
7. Fogaras, D., Racz, B.: Scaling link-based similarity search. In: WWW 2005, Chiba, Japan
8. Lizorkin, D., Velikhov, P., Grinev, M.: Accuracy estimate and optimization techmiques for SimRank computation. VLDB J. **19**(1), 45–66 (2010)
9. Jeh, G., Widom, J.: Simrank: a measure of structural-context similarity. In: KDD 02: Proceedings of the Eighth ACM SIGKDD International Conference on Knowledge Discovery and Data Mining, pp. 538–543. ACM Press, New York (2002)
10. Li, C., Han, J., He, G.: Fast computation of SimRank for static and dynamic information networks. In: EDBT 2010, 22–26 March 2010, Lausanne, Switzerland
11. He, G., Feng, H., Li, C.: Parallel simrank computation on large graphs with iterative aggregation. In: Proceedings of the 16th ACM SIGKDD 2010
12. Feng, H.: Research on Parallel Simrank. BeiJing Renmin University of China (2010)
13. Fogaras, D., Rácz, B.: Towards scaling fully personalized pageRank. In: Leonardi, S. (ed.) WAW 2004. LNCS, vol. 3243, pp. 105–117. Springer, Heidelberg (2004)
14. Yu, W., Lin, X., Le, J.: Taming computational complexity: efficient and parallel SimRank optimizations on undirected graphs. In: Chen, L., Tang, C., Yang, J., Gao, Y. (eds.) WAIM 2010. LNCS, vol. 6184, pp. 280–296. Springer, Heidelberg (2010)
15. Li, P., Liu, H., et al.: Fast single-pair SimRank computation. In: 2010 SIAM International Conference on Data Mining, pp. 571–582 (2010)
16. Langville, A.N., Meyer, C.D.: Updating pagerank with iterative aggregation. In: WWW Alt. 04: Proceedings of the 13th International World Wide Web Conference on Alternate Track Papers & Posters, pp. 392–393. ACM, New York (2004)
17. Page, L., Brin, S., Motwani, R., Winograd, T.: The pagerank citation ranking: bringing order to the web.Technical report, Stanford University Database Group. http://citeseer.nj.nec.com/368196.html (1998)
18. http://snap.stanford.edu/data/

OIPBP: On-Demand Interdomain Path Building Protocol

Yansheng Qu[1(✉)], Liang Li[1], Li Yan[1], Bo Mao[2],
and Xiaojuan Zhang[3]

[1]Shan Dong Electronic Power Corporation, Jinan, People's Republic of China
yansqu@foxmail.com
[2]Jiangsu Provincial Key Laboratory of E-Business,
Nanjing University of Finance and Economics, Nanjing, People's Republic of China
[3]School of Computer Science, Qinghai Normal University, Xining,
People's Republic of China

Abstract. Ill structured interdomain routing protocol makes interdomain route selection uncontrollable. To solve this problem, an on-demand interdomain path building protocol (OIPBP) is proposed in this paper. The main characteristic of OIPBP is that routers can lay control over the path selection process of their downstream nodes and customize the routes according to their own requirements. In order to achieve this goal, we extend BGP by adding more policies options into routing advertisements, and the inserted policies can be referred by the intermediate nodes when selecting paths. We verify OIPBP has good performance by experiments.

Keywords: Interdomain routing · OIPBP · BGP · Path selection

1 Introduction

There are two reasons that make interdomain routing uncontrollable. First, ASes can only get the routes advertised by other ASes. BGP is taken as the de facto interdomain routing protocol for the Internet. BGP gives each AS significant flexibility in deciding which routes to select and export. However, the available routes depend on the composition of the local policies in the downstream ASes, limiting the control each AS has over path selection. Second, in the intradomain routing, the management complexity of the networks arises mainly due to the interaction of ever increasing functionalities, and their required state information, with the distributed nature of routing design, in which each router independently computes and maintains the state required for its operation. Some research has been done to improve the controllability of Internet routing in recent years. For improving the controllability of interdomain routing, source routing [1–3], overlay networks [4] and multi-path routing [5–8] are proposed. For improving the controllability of intradomain, routing logically centralized intradomain control fashion such as RCP [9, 10] and 4D [11, 12] are proposed and studied in recent years. This paper is our initial effort in improving the controllability of interdomain routing of the Internet.

S. Zhou and Z. Wu (Eds.): ADMA 2012 Workshops, CCIS 387, pp. 41–51, 2013.
DOI: 10.1007/978-3-642-41629-3_4, © Springer-Verlag Berlin Heidelberg 2013

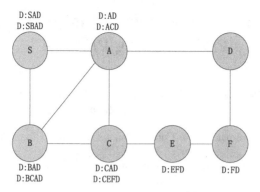

Fig. 1. BGP routing

Internet consists of thousands of independently administered domains (or Autonomous Systems) that rely on the Border Gateway Protocol (BGP) to learn how to reach remote destinations. Although BGP allows ASes to apply a wide range of routing policies, the protocol requires each router to select a single "best" route for each destination prefix from the routes advertised by its neighbors. This leaves many ASes with little control over the paths that the traffic takes. For example, an AS might want to avoid paths traversing a bad performance AS. Figure 1 is an example of this situation, where the source node S hopes to have a primary path SAD and another backup path not traversing node A for high reliability to destination D. These paths should exist since SAD and SBCEFD don't conflict with any network policy, however, S can't receive all these two paths because CEFD is filtered in node C by BGP's shortest-path rule.

Recent researches have considered several alternatives ways to interdomain routing over BGP, including source routing, overlay networks and interdomain multipath routing. In source routing, stub ASes have complete topology of the whole network and picks the entire path that packets traverse [1–3]. In overlay networks, packets can travel through intermediate hosts to avoid performance or reliability problems on the direct path [4]. The main idea of multi-path routing is to distribute more routes information, the way to achieve this goal is keeping BGP unchanged and extending BGP to transmit multi paths for one destination address. However, these techniques all have obvious shortcomings. The lack of control for ISPs is a significant impediment to the eventual adoption of source routing. In addition, both source routing and overlay networks may not scale to the size of the Internet. Current multipath proposals is limited to backup [8] or the success rate is unsatisfying. Instead, we explore a solution where the downstream nodes are capable of control the upstream ASes to select routes according to their requirements.

For simplicity, in the rest of the paper we use node representing AS of Internet, source node denoting the packets sender and also the starting point of the path, destination node denoting the receiver of packets and also the terminal of the path. Our solution is motivated by the fact about today's interdomain routing system: (1) At this stage, the routes provided by BGP are enough for most nodes in the Internet; (2) A small part of nodes has the diverse performance requirements and each router selects and advertises a single route for each prefix is not flexible enough. The main reason of the unsatisfying fact is that the downstream nodes just transmit the "shortest" path, which can't always represent the requirements of upstream nodes. This paper hopes to extend current BGP to a new protocol in which the upstream nodes are capable of laying control over the path selection of downstream nodes. Since current BGP can satisfy most nodes' need, we do not hope to change the logic framework of BGP much. Because the available routes of upstream nodes are decided by the downstream nodes in BGP, if the upstream nodes hope the downstream nodes to select paths according to their own needs, the requirements of the upstream nodes must be submitted to the downstream nodes before path selection.

The main idea of this paper is as follows: BGP is used to ensure the reachability of the source nodes and destination nodes. Based on this any source node that needs to build special paths sends the request to the corresponding destination node and the destination node starts an extra network convergence process to help the source build the satisfying paths. Our main contributions are as follows: (1) we extend BGP to a new protocol called P-BGP (Policy-based BGP) whose significant characteristic is more policy options are added into the routing advertisements. (2) we design OIPBP based on iterative P-BGP convergence process. The experimental results demonstrate the effectiveness of the proposed scheme.

The rest of this paper is structured as follows. In Sect. 2, we present related work. In Sect. 3 we discuss P-BGP and its implementation. In Sect. 4 OIPBP and how to extend current BGP to OIPBP are proposed. Section 5 the experiments and the results are discussed. Section 6 concludes our work.

2   Related Work

In this section, we present an overview of currently proposed alternatives to BGP. In source routing, each end host has complete knowledge of the entire topology and can choose whatever paths it wishes. In overlay networks, several overlay nodes to form a virtual topology on top of the physical network; each node can direct traffic through other overlay nodes en route to the destination. In interdomain multi path routing, some special paths are built for forwarding or backup along with the primary path.

2.1   Source Routing

In the past few years, several researchers have proposed source routing as a way to provide greater flexibility in path selection [1–3]. In source routing, the end hosts or edge routers select the end-to-end paths to the destinations. The data packets carry a list of the hops in the path, or flow identifiers that indicate how intermediate routers should direct the traffic. However, several difficult challenges remain to be solved before source routing can be deployed at scale:

- Limited control for intermediate ASes: Under source routing, intermediate ASes have very little control over how traffic enters and leaves their networks. This makes it difficult for intermediate ASes to select routes based on their local policies.

- Scalability: Source routing depends on knowledge of the network topology for sources to compute the paths, so the volume of topology data would be high. In addition, the sources must receive new topology information quickly when link or router failures make the old paths invalid.

Even source routing improves the controllability of the routing for the end hosts, intermediate ASes losing control over the traffic makes it difficult to deploy. We believe that it is valuable to consider other approaches that make different trade-offs between flexibility for the sources, control for the intermediate ASes, and scalability of the overall system.

2.2   Overlay Networks

In overlay networks, several end hosts form a virtual topology on top of the existing Internet [4]. When the direct path through the underlying network has performance or reliability problems, the sending node can direct traffic through an intermediate node. Although overlay networks are useful for circumventing problems along the direct path, they are not a panacea for supporting flexible path selection at scale, for several reasons:

- Data-plane overhead: Sending traffic through an intermediate host increases latency, and consumes bandwidth on the edge link in and out of that host. In addition, the data packets must be encapsulated to direct traffic through the host, which consumes extra bandwidth in the underlying network.
- Probing overhead: To compensate for poor visibility into the underlying network, overlay networks normally rely on aggressive probing to infer properties of the paths between nodes. Probing has inherent inaccuracies and does not scale well to large deployments.

Although overlays undoubtedly have an important role to play in enabling new services and adapting to application requirements, we believe the underlying network should have native support for more flexible path selection to support diverse performance and security requirements efficiently, and at scale.

2.3   Multi-Path Interdomain Routing

Recently some researchers proposed to build multi-path routing directly on the underlying network to improve the performance of routing [5–8]. Generally there are two schemes: keeping BGP unchanged and extending BGP to transmit more than one path for one destination. MIRO and D-BGP are the typical examples of these two schemes respectively.

MIRO [5] proposes to keep BGP unchanged and any node who needs an extra special path can build a tunnel with the destination node by negotiating. The built tunnel can be used for concurrent forwarding or backup. The biggest advantage of MIRO is that it keeps BGP unchanged and can be deployed incrementally. However, MIRO can't promise to find the needed path for the source node.

Different from MIRO, D-BGP [8] extends BGP to allow each router to advertise a most disjoint alternative path along with the best path. D-BGP provides fast failure recovery ability but it improves the size of routing table and the extra path can only be used for backup.

3 P-BGP

In this section, we extend BGP to a new protocol P-BGP (Policy-based BGP). The main difference between P-BGP and BGP is that more policy options called S_P options are added into routing advertisements, the added policies can be referred by the intermediate nodes when selecting paths. There are two kinds of S_P policies are added into routing advertisements in P-BGP: path selecting policy and path cancel policy. The path selecting policies are suggestions for the intermediate nodes on how to choose "best" path. And the path cancel policies direct the nodes to delete paths selectively.

The first three items of Table 1 are path selecting polices that can be added into path announcement advertisement. The last item of Table 2 is the path cancel policy that can be added into path cancel advertisement. We use four common policies temporarily and in the future research more policies can be added.

P-BGP is different from BGP on how to process path advertisements because more policies are referred in the decision-making process. In BGP, intermediate ASes select a "best" path according to local policies and the "shortest" rule from the candidates. However, in P-BGP, The "best" path is decided not only by the local policies of intermediate ASes but also by the S_P policies contained in the advertisements. The priority of the S_P policies in the path selecting process is lower than local policies and higher than the "shortest" rule. So the order the policies are referred is local policies, S_P policies and at last the "shortest" rule.

Table 1. S_P policies in routing advertisement in P-BGP

Policy	Explanation
Avoid(Q)	Avoiding paths traversing Q, Q is a node or link
Prefer(Q)	Prefer paths traversing Q, Q is a node or link
Disjoint(Q)	Prefer the path disjoint from Q, Q is a path
Except (Di,P1,P2,...Pn)	Cancel paths to destination address Di except P1,P2...Pn, P1,P2...Pn are paths

Table 2. Attributes of the data sets

Year	#of Nodes	#of Edges	P/C links	Peering links	Sibling links
2005	21392	47930	43184	3945	801
2007	26350	53178	47913	4370	895
2009	33670	62160	56158	4979	1023

We take the "avoid (Q)" policy for example. In Fig. 2, if node D generates an path announcement with the destination address D containing the policy "avoid(A)" and the announcement will arrive at node A directly and at C through F and E according to the network topology. Node A won't transmit any path to neighbors since the policy "avoid(A)" means the path traversing A will finally be neglected. If the local policy of node C is "Prefer(A)", then the remaining paths filtered by local policies is CEFD. CEFD does not conflict with the advertisement policy "avoid (A)" and survives.

4   OIPBP (On-Demand Interdomain Path Building Protocol)

In Sect. 3 we propose P-BGP whose main characteristic is that more S_P items are added into routing advertisements. In this section, we propose OIPBP based on P-BGP and describe its main process.

4.1   Main Idea of OIPBP

OIPBP is composed of two P-BGP converge processes: in the first converge process all routing advertisement are S_P policy items are empty and it is the same with BGP since P-BGP degrade to BGP if the S_P items of all advertisements are empty, the second is a P-BGP convergence process whose S_P options are not empty. The first convergence process is the foundation of the second. The main idea of OIPBP is as follows: The first P-BGP convergence process is used to ensure the reachability of the source node and the destination node; Based on the first convergence process, any source node who wants to build special paths can send the request to the destination and the destination node will generate a nickname for itself and start a global P-BGP convergence process taking the nickname as the destination address to help the source build the needed path, and the requirements of the new path are transformed into the S_P items of the advertisements in the second convergence process. When the second P-BGP process converges, the source node will receive the satisfied paths if they exist. As the by-products of the second process, many redundant paths are generated and another global convergence process is needed to eliminate them, where the cancel policies are added into the S_P options of the advertisements.

The algorithm of OIPBP is present as follows:

1. A P-BGP convergence process whose S_P policy items are empty. This convergence process is the same as current BGP since all the routing advertisements do not include S_P policies. After the P-BGP process converges, the source node and the destination node are reachable.
2. If any source node S wants to build a special path to the destination D, then S sends the request to D and along with the request the policies required for the new path are sent.
3. When D receives the request from S, D generates a nickname D' for itself and sends D' back to S.

4. After sending D' to S, D generates a new P-BGP process with the destination address D', and transform the policies received from S into the S_P items of the routing advertisements.

5. The intermediate nodes select the "best" path according to local policies and the S_P policies contained in the advertisement and transmit it to neighbors. After the politic process with the destination address D' converges, S will get the satisfying paths if they exist.

6. S selects the satisfying path and informs D to eliminate the redundant paths in the network. D generates a path cancel advertisement and adds the cancel policy into the S_P option of the advertisement. When the path cancel advertisement is distributed into the whole network, the redundant paths in the network are eliminated and the satisfying paths are reserved.

7. Steps (2) to (6) can be repeated arbitrarily to build enough number of paths.

By steps (1) to (6), the source node S can receive the routes with destination address D' and since it has known D' is a nickname of D, S actually has the required path to D. Moreover, S may use the built path for any purpose not only for backup but also for concurrent forwarding.

4.2 - An Example

We take Fig. 2 for example to describe OIPCP intuitively. Figure 2a is the routing table state after the BGP converges, and the local policy of node C is that CEFD and CAD are all legal, and C selects CAD as the "best" path since it is shorter. Assuming that S needs to send massive real-time data to node D and path SAD can't satisfy the requirements because of the limited capability of link A-D. S hopes to take another path which does not traverse link A-D for concurrent forwarding. However, the required path does not exists in S' routing table. We use OIPCP to build this path in the network.

The first phase of OIPCP is the same with BGP. By the path SAD built by the empty_SP P-BGP, S can communicate with D and request D to help him build a new path which does not traverse link A-D. S sends D the policies "avoid (A-D)" which means that S wants to build a new path to the destination D and dose not traverse the link A-D. When receiving the request of S, D generates a new nickname D' for itself

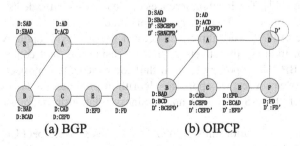

(a) BGP   (b) OIPCP

Fig. 2. BGP and OIPCP

and sends D' to S. After sending S the nickname, D generates a path announcement with the destination address D' and adds the policy "avoid(A-D)" into the S_P options of the advertisement. When A receives the advertisement, he does not transmit any path since there are no remaining paths filtered by the three steps: local policies, S_P policies and the "shortest" rule. When C receives the advertisement, path "CEFD" is selected after the three steps of filtering. When the politic P-BGP process converges, the network state is like Fig. 2b and S receives the path SBCEFD' to D' which does not traverse node A. Because S has known D' is a nickname of D, S actually has a satisfying path to D.

Because we take a global mechanism to help the source node build the satisfying path, they are many redundant paths generated in this process. After S receives the required path, S notifies D to eliminate the redundant paths in the network. So S informs D that path P' = SBCEFD' is hoped to be kept and other paths to D' be deleted. D generates a path cancel advertisement and transmits it to neighbors, and the policy except(P') is added into the S_P options of the advertisement. As the advertisement is distributed globally, all the nodes in the network delete the other paths to D' except P'. When this process converges, redundant paths in the network are deleted.

4.3   Multi-Path Interdomain Routing Implementation of OIPCP

Current Internet must be changed in several aspects to support OIPCP. First, the format of IP packet needs to be changed. Since nickname can be used as the destination address, the IP packet must be changed to support nickname and the destination item needs more bits. Second, the format of BGP advertisement must be changed and the S_P options could be added into it. In our current version, the advertisement just includes one S_P option. And in the future research, the advertisement should be designed to support dynamic number of S_P options.

5   Experiments

Ideally, we evaluate OIPBP by deploying it in the Internet and measuring the results. Since this is not possible, we simulate OIPBP operating in an environment as close to the current Internet as possible. Instead, we evaluate OIPBP on the AS-level topology, assuming that each AS selects and exports routes based on the business relationships with its neighbors [13]. The local preferences of the routes are decided solely based on AS relationships, and each AS is treated as one node.

We evaluate OIPCP under three instances of the AS-level topology, from 2005, 2007, and 2009, to study the effects of the increasing size and connectivity of the Internet on OIPBP. The topology is from the RouteViews [14] project. To infer the relationships between ASes, we apply the algorithms presented by Gao [15]. The key characteristics of the AS topology and business relationships are summarized in Table 2.

After inferring the relationships of the ASes, we take the Respect Export Policy in [8] as the local policy of every node, which means that the responding AS announces

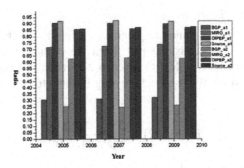

Fig. 3. Avoiding nodes in the network

all alternate routes that are consistent with the export policy. For example, an AS would announce all alternate routes to its customers, and all customer-learned routes to its peers and providers.

A. *Avoiding Nodes*

We take a horizontal comparing experiment to study the success rate of avoiding nodes in BGP, Source Routing, MIRO and OIPBP. MIRO is set like Ref. [5]. The result of this experiment is given in Fig. 3, in which the postfix a1 and a2 represent avoiding one node and avoiding two nodes in the network respectively.

Figure 3 shows that OIPBP has the similar success rate as Source Routing and it is far more than MIRO whether avoiding one node or two. Especially, when more nodes are needed to be avoided, the success rate of MIRO decreases sharply, but OIPBP maintains a flat downward trend. We think this is up to the link degree distribution in the Internet. In Internet, few nodes have the majority of link degrees of the network. Most paths of BGP traverse the nodes with big degrees so once the source want to avoiding these nodes, it is hard to find the satisfying path by negotiating in MIRO.

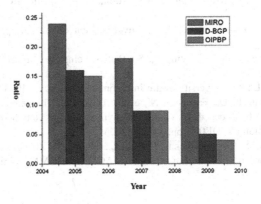

Fig. 4. Selecting a most disjoint path

B. *Selecting Most Disjoint Path*

Most disjoint path is a common requirement in Internet as the backup path. We implement building the most disjoint path from the primary path of BGP both in MIRO and OIPBP. The hops of MIRO is set to 2-3 in Ref. [5] and in this paper we relax the restrictions to 10. We use the following formula to calculate the similarities of two paths:

$$sim(P1, P2) = \frac{|P1 \cap P2|}{|\max(P1, P2)|}$$

The result is shown in Fig. 4.

Figure 4 shows that even the hops of MIRO are set to be very big, the success rate of MIRO is worse than OIPBP in selecting a most disjoint path from the basic path of BGP. And the performance of D-BGP and OIPCP are similar. However, the performance of MIRO and OIPBP and D-BGP has all greatly improved as the Internet grows.

Acknowledgement. This research is supported by National Natural Science Foundation of China (Nos.41201486, 71072172 and 61103229) and basic application research project of Qinghai science and technology department (NO. 2011-Z-719).

References

1. Argyraki, K., Cheriton, D.R.: Loose source routing as a mechanism for traffic policies. In: Proceedings of the Future Directions in Network Architecture (2004)
2. Yang, X.: NIRA: a new Internet routing architecture. In: Proceedings of the Future Directions in Network Architecture (2003)
3. Godfrey, P.B., Ganichev, I., Shenker, S., Stoica, I.: Pathlet routing. In: SIGCOMM, August 2009
4. Andersen, D., Balakrishnan, H., Kaashoek, F., Morris, R.: Resilient overlay networks. In: Proceedings of the SOSP (2001)
5. Xu, W., Rexford, J.: MIRO: multi-path interdomain routing. In: SIGCOMM, September 2006
6. Walton, D.A.R., Chen, E., Scudder, J.: Advertisement of multiple paths in BGP. draft-walton-bgp-add-paths-06, March 2006
7. Motiwala, M., Feamster, N., Vempala, S.: Path splicing. In: SIGCOMM, Seattle, WA, August 2008
8. Wang, F., Gao, L.: Path diversity aware interdomain routing. In: INFOCOM (2009)
9. Caesar, M., Caldwell, D., Feamster, N., et al.: Design and implementation of a routing control platform. In: Proceedings of the 2nd Symposium on Networked Systems Design and Implementation (NSDI'05), pp. 15–28 (2005)
10. Peterson, H., Sen, S., Chandrashekar, J., Gao, L., Guerin, R., Zhang, Z.: Message-efficient dissemination for loop-free centralized routing. ACM SIGCOMM Comput. Commun. Rev. **38**(3), 65–74 (2008)

11. Greenberg, A., Hjalmtysson, G., Maltz, D.A., et al.: A clean slate 4D approach to network control and management. ACM SIGCOMM Comput. Commun. Rev. **35**(5), 41–54 (2005)
12. Casado, M., Freedman, M.J., Pettit, J., Luo, J., McKeown, N., Shenker, S.: Ethane: taking control of the enterprise. In: SIGCOMM'07 (2007)
13. Caesar, M., Rexford, J.: BGP routing policies in ISP networks. IEEE Netw. Mag. **19**, 5–11 (2005)
14. University of Oregon Route Views Project. http://www.routeviews.org
15. Gao, L.: On inferring autonomous system relationships in the Internet. IEEE/ACM Trans. Netw. **9**(6), 733–745 (2001)

Social Web

An Automatic Spell Checker Framework
for Malay Language Blogs

Surayaini Binti Basri, Rayner Alfred$^{(\boxtimes)}$, Chin Kim On,
and Mohd Norhisham Razali

School of Engineering and Information Technology, Universiti Malaysia Sabah, Jalan UMS
88400 Kota Kinabalu, Sabah, Malaysia
surayaini_basri@yahoo.com, ralfred@ums.edu.my,
kimonchin@ums.edu.my, hishamrz@ums.edu.my

Abstract. A Spell checker is a system that is used to detect and correct misspelled word. Misspelled word is a word that exists in the existing lexicon that is not correctly spelled or in shortened form. These misspelled words often result in ineffective results of the Information Retrieval (IR) application such as document retrieval. This is because IR application should be able to recognize all words in a particular language in order to be more robust. The current spell checker for the Malay language uses a dictionary that contains pair of commonly misspelled word and its correctly spelled word in detecting and correcting misspelled word. However, this type of spell checker can only correct misspelled words that exist in the existing dictionary; otherwise it requires user interaction to correct it manually. This approach works well if the spell checker is a standalone system but it is not really an effective system when the spell checker is part of another IR application such as document retrieval for weblog. This is because there will be always new misspelled words created along with the increasing number of weblog pages. Thus, the number of misspelled words will also grow extremely. In this paper, we propose a new spell checker that detects and automatically corrects misspelled words in Malay without any interaction from the user. The proposed approach is evaluated by using texts that are selected randomly from the popular Malay blog. Based on the experimental results obtained, the proposed approach is found to be effective in detecting and correcting the Malay misspelled word automatically.

Keywords: Malay weblog · Information retrieval · Misspelled word · Spell checker

1 Introduction

The number of weblog pages written in Malay languages is increasing every day. Malaysian online users also constitute a large number of blogs in the blogs statistics. There are approximately 6.3 million Malaysian bloggers [7]. From this bloggers, Colette estimated that 46 % of Malaysia online users have their own blog pages [9]. Mishne also has stated that one of the main features of weblogs that differs from the common web page is the linguistic properties of blog content [5]. This is because a

S. Zhou and Z. Wu (Eds.): ADMA 2012 Workshops, CCIS 387, pp. 55–64, 2013.
DOI: 10.1007/978-3-642-41629-3_5, © Springer-Verlag Berlin Heidelberg 2013

weblog page is written by personal individual, so the word written in the blog can be in an informal format or containing typographic errors since there is no restriction for bloggers to write their blogs in a standard format. Janssen also stated that even in thoroughly corrected corpora like newspaper, typographic error also does occur [10]. This fact shows that the possibility of having misspelled words in personal weblog is much higher. According to Cambridge Dictionary Online [12], a misspelled word is a word that fails to be spelled correctly. These misspelled words may be formed from short form word, error in typing, or can be influenced by the dialect of the language. Information retrieval in weblogs becomes more challenging due to the existence of misspelled words in weblog pages. In order to retrieve information more effectively and efficiently, a better misspelled approach is required for Malay weblog.

Spell checker is a process of identifying misspelled word and convert it into its actual word. There are three types of spell checker that include post-processing spell checker, background spell checker and automatic spell checker [8]. The first spell checker is called a post-processing spell checker. The post-processing spell checker is a spell checker that is manually invoked by user after all or parts of the document are typed. This type of spell checker uses dictionary containing correctly spelled word. A word is identified as a misspelled word if that word does not match with any of the words listed in the dictionary. A list of correctly spelled alternative word then will be provided to the user via an interface such as dialog box. User can then choose any words from the list to correct the misspelled word or ignore the correction if they want the misspelled word remains uncorrected.

The second type of spell checker is called a background spell checker. The background spell checker detects a misspelled word as soon as online user enters the misspelled word but it is not automatically corrected. It still needs to invoke a post processing spell checker in order to correct the misspelled word.

The last type of spell checker is called an automatic spell checker that automatically detects and corrects misspelled word. It uses a dictionary containing pair of commonly misspelled word and its correctly-spelled word. This dictionary is called AutoCorrect dictionary. The misspelled word will be replaced automatically if the word entered by the user exists in the dictionary without any user interaction. However, this type of spell checker can only correct the misspelled word that exists in the dictionary. A post-processing spell checker still needs to be invoked if the misspelled word does not exist in the dictionary. This disadvantage can be solved if the list of Auto-Correct dictionary can be updated automatically.

Both the post-processing spell checker and the background spell checker perform well when it acts as a single system but unfortunately, it is not really practical for a spell checker that acts as a sub module for another IR application such as document retrieval for weblogs. This is because there will always be new misspelled words created along with the increasing number of weblog pages. Thus, the number of misspelled words will also grow extremely.' Handling every misspelled word that exists in a document manually is time consuming. Thus an automatic spell checker is needed to assist an information retrieval application to process a large number of documents more effectively and efficiently.

In this paper, we propose an approach to reduce the number of misspelled words that are unknown by the current lexicon. Section 2 describes some of the background related to misspelled checker approaches. Section 3 presents the proposed approach to reduce the number of misspelled for Malay weblog. Section 4 describes and discusses the experimental design and results. Section 5 concludes this paper.

2 Related Works

There are various ways of identifying and correcting misspelled words introduced by researchers for English articles. Travis introduced a post-processing spell checker to correct misspelled word [2]. In this approach, they use a dictionary called 'Corrected Before' file that contains a list of misspelled words and its correct words. Whenever a misspelled word is encountered and it is listed in the 'Corrected Before' file, it will be replaced with the respective corrected word, but if the word is not listed in the 'Corrected Before' file, the system will prompt the user with several options. These options are provided to correct the spelling, or possibly used as a tool to update the dictionary, assuming that the spelling of the misspelled word is correct. If the user selects to correct the misspelled word, then the system will prompt the user to input the correct spelling for the misspelled word. The system also will ask the user whether the correction should be added to the 'Corrected Before' file. If the user selects the option not to make any corrections, then the system will just correct the misspelled word without adding it to the 'Corrected Before' file else, the correction will be added to the 'Corrected Before' file before it corrects the misspelled word.

Walfish et al. have introduced an approach to automatically detect and correct a misspelled word in English language [8]. This approach applies several rules in finding the best replacement word for the misspelled word. For example, after selecting the suggested word which is generated for the misspelled word, they will compare the selected word with its misspelled word using a set of rules in finding the best replacement word for the misspelled word. If the misspelled word is not similar to the selected word based on certain criterion, the selected alternate words will be selected as the replacement candidate word. Selection criteria are then applied to the replacement candidate word and replacement indicator to get the replacement word. If the replacement indicator and the replacement candidate word satisfy the selection criteria, then one of the replacement candidate words will be selected as the replacement word.

Schabes and Roche have introduced an approach that is also able to automatically detect and correct misspelled words in English language [6]. The advantage of this approach is that the correction of the misspelled word is performed based on the surrounding context of the misspelled word. This approach applies the Finite State Machine (FSM) algorithm in detecting and correction the misspelled words.

In Malay language, there is not much research done to identify and correct mis-spelled words. However, there are some works related to processing unknown words in Malay done by Ranaivo Malancon Bali, Chong Chai Chua, Pek Kuan Ng in [15]. In their works, they identify and classify unknown words in Malay in which they have proposed an approach to identify abbreviation, proper name, loan word and affixed word for formal Malay text. Although the proposed approach performs well for formal Malay text, but it may produce a huge amount of errors in processing Malay weblog since most of the Malay weblogs are written in an informal language and not according to the spelling rules of formal Malay.

For example, one of the rules of identifying loan word is, if the word ends with any of these characters {e, o, c, j, w, y, ks, ans, oid, asma, isme, logi, grafi}, then it is considered as a loan word. Unfortunately, in Malay weblogs, some people will replace the word that ends with "a" to "e". For example, the word "saya" becomes "saye" due to because the Selangor dialect pronunciation.

Besides that, the paper also proposes that if a word contains any of these character sequence {ee, oo, uu, i.e., bb, cc, dd, hh, jj, ll, mm, pp, qq, rr, ss, tt, vv, ww, xx, yy, zz, ph} then it is also considered as a loan word. This is not applicable in Malay weblogs because some people will use a short Malay text form such as "sbb" for "sebab" or "ttp" for "tetap".

The process of identifying abbreviation is also not applicable in Malay blogs. For example, in their paper, it is proposed that a sequence of any consonant letters written in uppercase is considered as abbreviation. For instance "JKR" that usually stands for "Jabatan Kerja Raya". However, it is not applicable for the word used in a Malay blog because most people are not following any rules in writing their blogs. For example, they sometimes simply type any short term word in capital letter to show their feeling. For instance, "anda TDK dibenarkan melawat laman ini". The word "TDK" which is a short form of "tidak" in Malay and "No" in English in the sentence shows that the word is highlighted to enhance its meaning.

Another research conducted by Kasbon et al. that translates short form texting language to its complete and correct forms [4]. In their works, the proposed approach checks whether the sentences have any grammatical and structural errors. Then, it will suggest the type of correction to the sentences. This spell checker is an automatic spell checker for Malay articles. However, it also inherits the disadvantage of an automatic spell checker in which the detection and correction of the short form misspelled words only can be done for words that exist in their database. In other words, the system will not be able to detect and correct any misspelled words that are not listed in their database. As a result, a new robust method is required that is able to perform the spell checker that do not depend entirely on the database.

3   A Framework for a Malay Language Spell Checker

In this paper, we propose a spell checker approach that identifies and corrects mis-spelled words automatically. In our approach, there are five main modules that the system should go through in detecting and correcting a misspelled word. The first module is called the tokenization module, followed by a symbol elimination module. The third module is called the English word elimination module and the fourth module covers the stemming process and the last module performs the misspelled word identification. As illustrated in Fig. 1, these modules will be described in detail below.

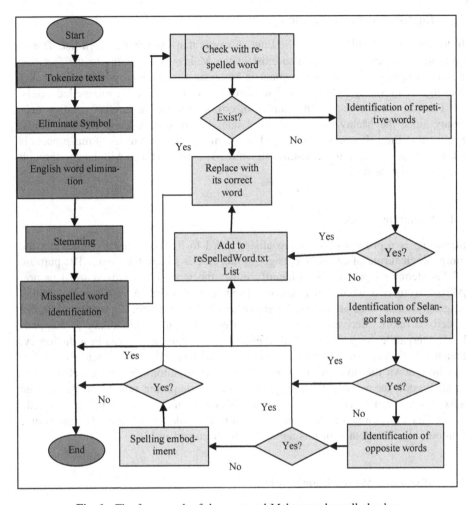

Fig. 1. The framework of the proposed Malay words spell checker

3.1   Tokenization

Firstly, all the texts extracted from a collection of weblog articles are tokenized and form a list of unique words. This word list later will go through the process of symbol elimination.

3.2   Elimination of Irrelevant Symbols

This is the step where all the symbols (except alphabet and word combination with number, full stop (".")) are eliminated such as '?', '!', '#', '$', '%', '^', '*', '~'. The full stop symbol is needed in order to identify the sentence of the word. The word list without symbols will then go through the process of English word elimination.

3.3   English Words Elimination

In this process, we filter out all the English words that exist in the word list. This is because many English words also exist in the Malay weblogs due to the fact that English language is the second language in Malaysia. Since the proposed spell checker is designed for Malay weblogs, English words will be considered as misspelled words.

Thus, in order to reduce the number of errors, eliminating English words is necessary since the Malay misspelled word does not recognize these English words. Eliminating English words requires an English dictionary. The result of this process is a list of words that contains non-English words. The list produced will then go through the process of stemming.

3.4   Stemming Process

Stemming is a process of reducing affixed word to its root word by following the morphological rule of certain language (in this case is Malay language). The purpose of this stemming process is to identify either the word is an affixed word or not. Affixed word is a word that is derived from its root word. A root word of Malay language can produce about five derivative words. For instance, if a root word "jadi" is derived into its derivative words, we can get "menjadi", "dijadikan", "menjadi-kan", "terjadi", "terjadinya". If all the derivative words are stored in a dictionary, then the dictionary size will also increase five times larger than the current size, thus it can slow down the process of stemming. In order to automatically extract the misspelled word, the word list has to go through the stemming process so that all the affixed words can be tagged as known words and the unknown words can be listed.

We have developed a stemmer that uses root words from Kamus Dewan Fourth Edition [11]. The details of this approach are discussed in [3].

3.5   Misspelled Words Identification

One of the causes of a misspelled word is typo word. Typo word is a word that exists in Malay language but the word is spelled based on their pronunciation. These words are not recognized by any systems because they are not spelled according to the standard Malay spelling rules. Since there is no research done in identifying these typo words in Malay, it requires analyzing features of the typo words in Malay blog. We have to know several features of the typo word so it can be referred during the development of the approach in identifying typo words later. Since our scope is only to identify misspelled words in Malay Selangor pronunciation, so all features discussed in this paper will refer to the Malay Selangor pronunciation. These features include the repetitive words, Selangor pronunciation words and opposite words. These features are the common features observable in most Malay weblogs.

The process in this module starts when the remaining unknown words from the stemming process is going through the process of matching the misspelled word with its correctly-spelled word. The system will check if the misspelled word exists or not in the preliminary dictionary word called 'reSpelledWord.txt' that contains misspelled words and its correctly-spelled word. If the word exists in the dictionary then the word will be replaced with its correctly spelled word. If the word does not exist, then the misspelled word will go through the process of identifying repetitive words. These repetitive words are identified based on the character '2' that is used at the end of the word. This is because, in most Malay blog, a character '2' is often used to indicate a repetitive word. For example, "macam2" is a short form of "macam-macam" (variation). In the system, if the word ends with '2', then it will remove the character '2' and recheck either the word without '2' exists in the dictionary or not. If it exists, then the word "macam" will be corrected by repeating it two times with hypen (-) in the middle to become "macam-macam". The "macam2" and its correction then will be added to the "reSpellWords.txt" file so that every time the word "macam2" exists, it will be automatically replaced with its correctly-spelled word without going through all the process again. If the word without '2' does not exist in the dictionary or does not ends with '2' then it will then be processed in the next stage which is the process of identifying Selangor pronunciation. Selangor pronunciation is identified based on the last character of the word. If the word ends with a character 'e', then the system will replace the character 'e' with 'a' because based on Selangor pronunciation, they always replace any words that end with a character 'a' with 'e'. For example, the word "saya" ("I" in English) becomes "saye" or "suka" (like) becomes "suke".

After the replacement, the system once again will check if the word exists in the dictionary or not. If it exists in the dictionary, then the misspelled word and it correction will be added to the "reSpelledWord.txt" file, so that every time this misspelled word is encountered, it will be automatically replaced with its correctly-spelled word without going through the same process again. If the word does not exist in the dictionary or if the word does not end with 'e', then it will go through the process identifying opposite words. Opposite words are identified by the character 'x' at the first character of the word. Bloggers always rewrite "tidak" (no) with a character 'x'. For example "xsuka" means "tidak suka" (do not like) or "xcantik" means "tidak cantik" (not beautiful/ugly). If the word start with a character 'x' then the system will remove the 'x' and recheck the word against the dictionary.

If the word exists, then the word will be added with 'tidak' before the word and will be added to the 'reSpelledWord.txt' file, so that every time this word is encountered again, it will be automatically replaced with its correctly-spelled word without going through all the process again.

If the word doesn't match with any features mentioned above then it will go through the process of spelling embodiment. In the spelling embodiment process, a list of alternative words will be provided to the remaining unknown word from previous process. To do this, we apply the existing spell checker API for java called Jazzy [13]. One of the functions that can be applied in Jazzy is to provide a list of alternative words for the misspelled word by inserting, deleting, replacing or transposing characters in the misspelled word until the correctly-spelled word is obtained.

After the list of the alternative words is obtained, each of the alternative words then will be ranked to select the best candidate word for the misspelled word. The process of ranking is performed based on the Levenshtein distance formula which is also known as Edit distance formula [14]. Using this formula, the rank of each the alternative words can be determined based on the difference between the misspelled

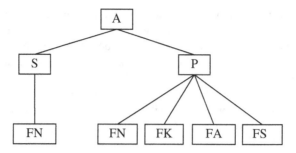

Fig. 2. Grammar structure of a sentence

word and the alternative word. The first three smallest value of difference between the misspelled word and the alternative word will be chosen. These three words will then be ranked based on the grammar structure of the sentences. There are four basic sentence patterns in Malay [4] (Fig. 2).

A sentence comprises of a subject (S) and a predicate (P). A subject is made up from noun phrase (FN) and a predicate can be made up from the combination of noun phrase (FN), verb phrase (FK), adjective phrase (FA) and preposition phrase (FS). For instance, the sentence "Ramli seorang yang rmah" contains a misspelled word "rmah". Since this word does not exist in the "respelled.txt" file, and it does not fulfill the criteria of repetitive word, Selangor slang or opposite word, then it will go through the process of spelling embodiment. In this process, several alternative words and its ranking are suggested. Let's say there are three alternative words suggested which are "ramah-1", "rumah-1", and "remah-1". All these alternative words having ranking 1 which means that, there is only 1 difference between the misspelled word and the alternative word. The grammatical information for these alternative words is needed in order to choose the most suitable word based on its part of speech tag in order to replace the misspelled word.

4   Experimental Design and Results

In order to evaluate our approach, we take the Malay weblogs written by Malaysian as our test datasets. The Malay weblogs have received some awards at the World Blogger and Social Media Awards 2012 [1]. This event was organized by Minggu Media Sosial Malaysia (MSMW) and launched at Putra World Trade Centre, PWTC on 16 March 2012. These weblogs are categorized into a few categories (e.g., education, celebrity, photographer, entertainment, business and etc.). In our experiment, only 11 out of 23 weblogs categories are used since the other 12 categories are written in English. These weblogs were chosen because these weblogs are considered as the most visited weblogs in Malaysia.

After eliminating all symbols from the datasets, we managed to get 11896 words. From these words there are 3046 distinct words obtained. After eliminating the English words, there are 2675 words left. We apply the spell checker for these 2675 distinct words and 2453 words are successfully identified and corrected. That means only 232 words or 8.67 % words are failed to be detected and corrected by using our proposed approach. Some of these 232 words (82 words) are proper name such as "nurhaliza" and "iphone". The other 150 words are constituted of unproductive words such as "somi" that actually refers to a word "suami" (husband), a Malay affixed plus English word such as "disuggestkan" (suggested) or "shootingnya" (his/her shooting), abbreviation words such as "vvip" or "psp", affixed words that are failed to be identified by the stemmer includes "menyayangiku"(loves me), loan words from Arabic language such as "hadith", and a combination of number and character such as "rm5", "tv3".

5 Conclusion

Based on the percentage of errors in the conducted experiment, it can be concluded that, the proposed approach is able to detect and correct a simple misspelled word automatically and effectively without any interaction from user. This approach can be improved further and used as part of any IR applications such as document retrieval by implementing some of the rules identified based on the 232 words that are failed to be detected and corrected. However, improvement still needs to be done in order to reduce the number of error such as finding a formula on how to rank the alternative words based on the context of the word. In order to do that, we may need a lexicon Malay dictionary that consists of Malay word and its type of category [4]. Besides that, we may add some rules in order to detect and correct the affixed English word, the loan word and also to improve the stemming process as already discussed in [3].

Acknowledgments. This work has been partly supported by the LRGS and RAGS projects funded by the Ministry of Higher Education (MoHE), Malaysia under Grants No. LRGS/TD/2011/UiTM/ICT/04 and RAG0008-TK-2012.

References

1. World Blogger and Social Media Awards 2012 (2012, March 16). http://www.socialmediaweek.com.my/awards/index.php. Accessed 18 July 2012
2. Travis, P.A.: Patent No. 5604897. United States of America (1997)
3. Leong, L.C., Basri, S., Alfred, R.: Enhancing Malay stemming algorithm with background knowledge. In: Anthony, P., Ishizuka, M., Lukose, D. (eds.) PRICAI 2012. LNCS (LNAI), vol. 7458, pp. 753–758. Springer, Heidelberg (2012)
4. Kasbon, R., Amran, N., Mazlan, E., Mahamad, S.: Malay language sentence checker. World Appl. Sci. J. (Special Issue on Computer Applications and Knowledge Management) **12**, 19–25 (2011)
5. Mishne, G.: Information access challenges in the Blogspace. 1 (2007)
6. Schabes, Y., Roche, E.: Patent No. US 7853874 B2. United States of America (2010)
7. Ulicny, B.: Modeling Malaysian public opinion by mining the Malaysian Blogosphere. 5 (2008)
8. Walfish, M., Hachamovitch, Andrew, F.: Patent No. 6047300. United States of America (2000)

9. Colette, M.: Blogging Phenomenon Sweeps Asia. MSN Press Release. 27 November 2006
10. Janssen, M.: Orthographic neologisms: selection criteria and semi-automatic detection. http://maarten.janssenweb.net/publications (unpublished)
11. Abdullah, M.T., Ahmad, F., Mahmod, R., Sembok, M.T.: Rules frequency order stemmer for Malay language. IJCSNS: Int. J. Comput. Sci. Network Secur. 9(2), 433–438 (2009)
12. Cambridge University Press (n.d.). Cambridge Dictionary Online. http://dictionary.cambridge.org/dictionary/british/misspell. Accessed 16 July 2012
13. Sourceforge (n.d.). The Java Open Source Spell Checker. http://jazzy.sourceforge.net/. Accessed 31 June 2012
14. Levenshtein, V.: Binary codes capable of correcting deletions, insertions, and reversals. Cybern. Control Theory 10(8), 707–710 (1966)
15. Bali, R.M., Chua, C.C., Ng, P.K.: Identification and classification of unknown words in Malay language

Finding and Extracting Academic Information from Conference Web Pages

Peng Wang[1(✉)], Xiang Zhang[1], and Fengbo Zhou[2]

[1]School of Computer Science and Engineering, Southeast University, Nanjing, China
{pwang@seu.edu.cn, x.zhang}@seu.edu.cn
[2]Focus Technology Co., Ltd, Nanjing, China
zhoufengbo@made-in-china.com

Abstract. This paper proposes a method for finding and extracting academic information from conference Web pages. The main contributions include: (1) A lightweight topic crawling method based on search engine is used to crawl academic conference Web pages. (2) An new vision-based page segmentation algorithm is proposed to improve the result of classical VIPS algorithm by introducing complete tree. This algorithm can divide Web pages into text blocks. (3) Using bayesian network classifier, all text blocks are classified as 10 categories according to its vision features, key-word features and text content features. The initial classification results have 75 % precision and 67 % recall. (4) The context information of text blocks are employed to repair and refine initial classification results, which are improved to 96 % precision and 98 % recall. Finally, academic information is easily extracted from the classified text blocks. Experimental results on real-world datasets show that our method is effective and efficient for finding and extracting academic information from conference Web pages.

Keywords: Topic crawler · Web information extraction · Page segmentation

1 Introduction

Current structural or semantic academic data such as ArnetMiner academic researcher social network [1], is based on database like DBLP and ACM library. These academic data mainly describes paper publication information of researchers. However, academic activity knowledge is not included by current academic data. Academic conferences websites not only contain paper information, but also contain many academic activity information, which includes research topic, conference time, location, participants, academic awards, and so on. Obtaining such information is not only useful for predicting research trends and analyzing academic social network, but also is the important supplement to current academic linked data. In order to automatically and efficiently obtain the clean and high quality academic data, it is necessary to extract useful academic information from these conference Web pages. Since academic conferences Web pages are usually semi-structured and have diversity content, there is no a unified way to automatically find conference sites and extract the academic information.

S. Zhou and Z. Wu (Eds.): ADMA 2012 Workshops, CCIS 387, pp. 65–79, 2013.
DOI: 10.1007/978-3-642-41629-3_6, © Springer-Verlag Berlin Heidelberg 2013

To find academic information, it needs to crawl thousands of conference websites automatically, that is a topic crawling problem. To avoid to check all Web pages one by one, we should design a lightweight topic crawler. Web information extraction is a classical problem [2, 3], which aims at identifying interested information from unstructured or semi-structured data in pages, and translates it to into a semantic clearer structure such as XML, relational data and domain ontology. Although academic conference pages usually have strict layout and content description, there is no a fixed template for all conference pages to follow, and the script language used to present content is also different. Therefore, it is difficult to find a general extraction model.

In this paper, we propose a method to find and extract useful academic information from conference Web pages automatically. First, the conference websites are collected by a lightweight crawling method. Then, given a sample conference Web page, it is segmented into a set of text blocks using an algorithm which combines vision-based and DOM-based segmentation methods. Third, text blocks are classified into pre-defined categories using bayesian network, in which each text block is represented by some features including vision features and semantic features. Post-processing can improve initial classification results by repairing wrong results and adding unclear results. Finally, we integrate the extracted information from a conference website to obtain the clean and high quality academic data.

In summary, this paper has following contributors: (1) A lightweight topic crawling method based on search engine is proposed. It collects academic conference sites through search engine by queries of initial seeds. Then it collects more conferences by links on these conference sites. Meanwhile, a filter is used to select relevant conferences. More seeds is obtained during the crawling by analyzing new conferences. (2) We propose a new vision-based page segmentation algorithm, which use DOM tree to compensate the information loss of classical vision-based segmentation algorithm VIPS. (3) We transform the conference Web information extraction problem into a classification problem, and classify text blocks as 10 pre-defined categories according to vision, key words, text and content information. (4) We improve the classification quality by post-processing. Our experimental results on the real world datasets show that the proposed method is highly effective and efficient for extracting academic information from conference Web pages.

2 Crawling Academic Conference Websites

Most Web pages are linked by hyperlink between them. For the websites in a domain, there are some Hubs and Authorities [4]. Hubs refer to the websites which have many hyperlinks to authority pages. Authorities are the websites that there are many hyperlinks link to them. In addition, in a special domain, the link density between pages are high than the link density between these pages and pages of other domains [5], namely, a Web community is a collection of Web pages in which each member page has more hyperlinks within the community than outside it. We still use hyperlink to crawl more websites in academic conference domain. Figure 1 shows our crawling model, which has five main steps as follows:

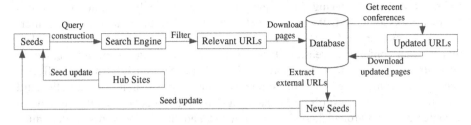

Fig. 1. The lightweight crawling model

(1) *Input processing*: Given some seeds, which contains short name and full name, we generate a series queries with year and run these queries by search engine, then we can obtain URLs of conferences.

(2) *Crawling and downloading*: We download pages according to URLs returned in step 1 and save pages into local database. Meanwhile, we extract links to external websites.

(3) *Filtering relevant pages*: For each external URL, we calculate its relevant score to determine whether it is a relevant site. If a URL is relevant, we will extract conference short name and full name, and add it as a new seed.

(4) *Hub conference pages*: If there is no seed, a manually maintained pages which contains a list of conference URLs is imported. New seeds can be obtained from these hub pages.

(5) *Crawling updating*: We will crawl obtained URLs periodically, the lasted conference pages will be re-crawled more quickly than older ones.

Crawling Seeds: The principle of our crawler is also employing the link between different conference websites. Namely, links to external websites can help us to find new conference websites. Our initial seeds only contain conference short names and full names. We select initial seeds from ArnetMiner conference rank list.[1] Here, we refer the top 150 conferences as rank A conferences, and the 151−500 conferences in the list are rank B conferences, then other conferences are rank C conferences. We randomly select some rank A, rank B and rank C conferences as initial seeds. A seed is a 2-tuple: <ShortName, FullName>. For a given seed, we can construct a series query key words like ShortName+Year, then use Google API to find the right conference websites. In the query key words, Year is decrease from 2012 with step 1. If we cannot find relevant conference websites in sequent 3 years, we stop to construct new query key words. For example, <IJCAI, International Joint Conference on Artificial Intelligence> is a seed, then IJCAI2012, IJCAI2011 and IJCAI2010 are the corresponding query key words. When a conference website is found, we analyze all external URLs. If an URL links to a new conference, we can add this conference into seed queue. Since we don't know the real link distribution between academic conference websites, initial seeds should cover conferences in different rank.

---

[1] http://arnetminer.org/page/conference-rank/html/Conference.html

Relevant Website Filtering: For a query key words, we only examine top-k results returned by Google API. We need to determine which result is an academic conference websites that we find. We use a three-level filtering method based on SVM to find relevant conference websites. First level is the preliminary topic filter based on URL string, short name and full name of a website. Second level uses a collected document of conference key words to filter websites. For remaining websites, the third level filter extracts key words of the main page, and then calculates similarity between a conference background knowledge document and these key words to determine whether the websites is relevant. This three-level filter can remove the unrelated websites efficiently.

Crawling Update Strategy: According to the fact that web pages update satisfy Poisson distribution, we calculate the average update periods of web pages in a website as its initial update period. Then we crawl a website periodically and adjust its crawling update period according to changes of Web pages.

Conference Hub Websites: To compensate for limitations of academic conference websites may lack of links between them, we maintain some conference hub websites, which contains a lot of academic conference list and corresponding external URLs. Once there is no seed to be crawled, we periodically visit these conference hub websites and add these conferences as new seeds.

3   Web Page Segmentation

To extract the academic information, we first segment Web pages into blocks by VIPS [6], which is a popular vision-based page segmentation algorithm. VIPS can use Web page structures and some vision features, such as background color, text font, text size and distance between text blocks, to segment a Web page. In VIPS, a Web page Ω is represented as a triple $\Omega = (O, \Phi, \delta)$. $O = (\Omega_1, \Omega_2, ..., \Omega_N)$ is a finite set of blocks. All blocks are not overlapped. Each block can be recursively viewed as a sub-Web-page associated with sub-structure induced from the whole page structure. $\Phi_i = (\varphi_1, \varphi_2, ..., \varphi_T)$ is a finite set of separators, including horizontal separators and vertical separators. δ is the relationship of every two blocks in O and can be expressed as: $\delta = O \times O \rightarrow \Phi \cup \{NULL\}$. Figure 2 shows the segmented results of ACL2011 main homepage, which is segmented into two blocks VB1 and VB2 with separator $\varphi 1$. VB1 and VB2 are then segmented into more small blocks such as VB1_1 and VB2_3. After three segmentations, we have the 10 text blocks. These text blocks can be constructed as a vision tree, which assures that all leaf nodes only contain text information.

VIPS can obtain good segmentation results for most Web pages, but we find it would lose some important information when deal with some Web pages. It is caused by the reason that VIPS algorithm is only based on vision features of page elements, so it would ignore blocks whose display is inconsistent. Therefore, we introduce DOM-based analysis to improve VIPS segmentation results, especially finding missed text blocks.

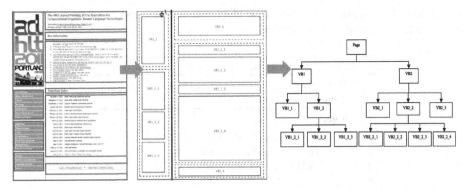

Fig. 2. An example of Web page segmentation by VIPS

First, we need to obtain the basic vision semantic blocks by analyzing DOM tree of Web pages. A vision semantic block is a text block with independent meaning. A lot of blank nodes are removed from HTML tags. Then we traverse DOM tree to extract vision semantic blocks. A vision semantic block is between two newline tags such as <br/> and only contains style tags and texts.

Algorithm 1 shows the detail of generating the VIPS complete tree. Let SB be vision semantic block, LN be layout node of vision tree generated by VIPS, and DN be data node. This algorithm includes three steps: (1) It finds a SB by traverse LN to search matched layout nodes; (2) If it finds a matched layout node, then this node is also a vision semantic block; (3) If it does not find a matched layout node, then add this SB into the vision tree. This algorithm not only assures that there is no information loss, but also preserves the structure of vision tree. Figure 3 shows the part of VIPS complete tree for bottom part of AAAI2010 main page. As vision tree shows, VIPS only extracts the text with bold font. It is not the result we expect. After the

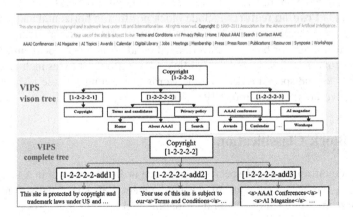

Fig. 3. An example of generating VIPS complete tree

processing of Algorithm 1, new semantic blocks are added to the vision tree, which is the complete tree with correct text blocks.

Algorithm 1: Generating VIPS complete tree algorithm
Input: <LayoutNode,DataNode> LNDN[], VIPS result PN
Output: a VIPS complete tree T

1	**begin**
2	**for** (PN.children[i] in PN.children[])
3	**if** (LNDN[] has key PN.children[i])
4	add LNDN[PN.children[i]] to T
5	**else**
6	add new DataNode(PN.children[i]) to T
7	LN saved[].add(PN.children[i])
8	**end**
9	**while** (LN saved[] is not empty) {
10	currentLN = LN saved[0]
11	LN saved[].remove(0)
12	currentDN = LNDN[currentLN]
13	**for** (currentLN.children[i] in currentLN.children[])
14	**if** (LNDN[] has key currentLN.children[i])
15	add LNDN[currentLN.children[i]] to T
16	**else**
17	add new DataNode(currentLN.children[i]) to T
18	LN saved[].push(currentLN.children[i])
19	**end**
20	**end**
21	**end**

Since some blocks such as navigation, copyright and advertisement do not contains the academic information. We regard these blocks as noise, which should be removed from VIPS complete tree. The noise removing process uses some vision features [7]. (1) **Position features** include block position in horizontal and vertical on page and ratio of block area to page area. (2) **Layout features** include alignment of blocks, whether neighbor blocks are overlapped or adjacent. (3) **Appearance features** include size font, image size, and font of link. (4) **Content features** include common words of blocks and special order of some words. According to these vision features, we can remove noise nodes from VIPS complete tree.

4   Text Block Classification

The academic information of a specific conference is distributed within a set of pages. For instance, the *Overview* page of the conference Web site usually contains the conference name, time, and location information, and a *Call for Papers* page usually contains topics of interest and submission information. In general, we are primarily concerned with five types of academic information on a conference website: (1)

Table 1. Categories of text blocks

	Category	Description
Date	DI(dateItem)	Dates about conference events
Location	PI(placeItem)	Location function and address
Research	AR(area)	Research area in high level
	TO(topic)	Research topic in each area
People	PO(position)	Positions of participants such as *Speaker*, *Chair* and *Co-Chair*
	PE(peopleItem)	People names and institutions
Paper	PA(paper)	Paper type, title and authors
Other	CO(collection)	Set of some above blocks, such as DI + PI means that it contains date and location. For example, a PI + DI in AAAI-11page: "AAAI is pleased to announce that the Twenty-Fifth Conference on Artificial Intelligence (AAAI-11) will be held in San Francisco, California at the Hyatt Regency San Francisco, from August 7–11, 2011"
	R(related)	Not belong to above 8 categories, but contains useful information, such as workshop information
	N(notRelated)	Not belong to above 8 categories and does not contain useful information

Information about conference date: conference begin and end date, submission deadline, notification date of accepted papers, and so on. (2) *Information about conference research topics*: call for papers (or workshops/research papers/industrial papers), topics of interests, sessions, tracks and so on. (3) *Information about related people and institute*: organizers, program committee, authors, companies, universities, countries and so on. (4) *Information about location*: conference location, hotel, city and country. (5) *Information about papers*: title, authors of papers.

We divide all text blocks into 10 categories as Table 1 shows: (1) **DI**: It describes date information; (2) **PI**: It describes location information; (3) **AR**: It refers to top level information such as research area; (4) **TO**: It refers to research topics, and a AR block may have some corresponding TO blocks; (5) **PO**: It describes the role of people in conference such as Speaker and Chair. (6) **PE**: It refers to information of a person; (7) **PA**: It is the information about papers; (8) **CO**: It refers the blocks which is combined by the above 7 categories blocks; (9) **R**: It refers to the interested blocks but not belong to any categories; (10) **N**: It refers to the blocks not only belong to any categories but also not related to academic information. Figure 4 shows each category and corresponding examples.

According to these categories, we can select some features to measure a given text blocks. We use vectors as Table 2 shows to describe each blocks. For a text block, we construct its features according to vision, key words and text content information. For example, given a text block: "Paper Submission Due: **Friday, May 6, 2011 (23:59 UTC - 11)**" and its HTML source code: *<li> Paper Submission Due: Friday, May 6, 2011 (23:59 UTC - 11) </li>*, its feature can be constructed as following.

(1) **Vision features**: isTitle = false, isHeader = false, startWithLi = true, left = (280-0)/950 = 0.3 (page width:950, left margin: 0, text left margin:280), with = 640/950 = 0.7 (text width: 640);

Fig. 4. Examples for text block categories

Table 2. Feature vectors of text blocks

Vector	Description	Value
isHeader	Whether the biggest font size	bool
isTitle	Whether the title font size	bool
nearestTitle	Type of the nearest title block	int
textLength	Length of text block	int
fontSizeToAverage	Average font size	int
fontWeightToAverage	Average font weight	int
startWithLi	Whether start with <li>	bool
dateTypeNum	Number of key words about date type, such as *deadline*	int
dateNum	Number of key words about date, such as *January*	int
placeTypeNum	Number of key words about location type, such as *Place*	int
placeNum	Number of key words about location, such as *Italy*	int
areaNum	Number of key words about research area	int
nameNum	Number of names	int
institutionNum	Number of institutions	int
positionNum	Number of positions	int
authorNum	Number of authors	int
abstractTypeNum	Number of key words about abstract	int
paperTypeNum	Number of key words about paper type	int
wordNum	Number of words of text blocks	int
wordToName	Ratio of number of words to number of names	double
linkTotext	Ratio of length of link to length of blocks	double
left	Ratio of left margin to page width	double
width	Ratio of block width to page width	double

(2) **Key word features**: nearestTitle = DI (its nearest and isTitle = true blocks is about date information), dateNum = 2 (it contains 2 date words: Submission and Due), paperTypeNum = 1 (it contains 1 key word about paper: Paper);

(3) **Text content features**: fontSize = 0, fontWeight = 0, textLength = 58, text-Link = 0, wordNum = 11, nameNum = 5, wordToName = 11/5 = 2.2.

There are many famous existing classification algorithms such as C4.5 [8], K-Nearest Neighbors (kNN) [9] and Bayes Network [10]. C4.5 and Bayesian Network are the most widely used classifier models. After comparing the two classifier models, we choose bayesian network model to solve the text blocks classifier problem.

5 Post-processing

This classified results can be improved by post-processing, which includes repairing wrong classified results and adding missed classified results. The text blocks with wrong classification can be determined by two sides:

(1) A block has special features but cannot be classified correctly. For example, given text block "*Camera Ready Papers Due: Thursday, August 11, 2011*", which have typical DI features, namely, *dateNum* = 1, *dateTypeNum* = 1, *isDateArea* = true and *isPreviousDate* = true. However, this block is classified as Related. For this situation, it can be repaired by identify some typical combination of features. This method is suitable for text blocks with clear features, such as DI, TO, PO and PE.

(2) A text block is classified as a category but it does not contain corresponding features. For example, text blocks about submission instruction would be classified as CO. Although it contains a lot of words, it usually does not contain any people name, date or location. Therefore, we can check its feature values *wordToName*, *dateNum*, *dateTypeNum* and *placeNum* to determine whether it is a CO category. According to this way, we can repair some wrong classifications.

Some text blocks are classified as R category, and they have useful information, but their categories are not clear. These blocks should be checked further. Therefore, we use context feature to determine the blocks with R category. The context feature consists of 11 boolean values: *isDateArea, isPlaceArea, isTopicArea, isPeopleArea, isPaperArea, isPreviousDate, isPreviousPlace, isPreviousTopic, isPreviousPeople, isPreviousPaper* and *isVisuallySame*. We propose some rules to add text blocks without clear classification.

Rule 1: DateItem complete rule : A DI text block usually appears as three situations: (1) It appears as page title with bold and big font size; (2) It appears with other DI text blocks; (3) It appears separately. Since a DI block is very dependent to key words, for any situation, key words and other features should be considered.

For situation (1), we can use a simple rule to detect: *dateNum*>0 && *wordNum*<=5 && *isTitle*=ture.

For situation (2), the corresponding rule is : (*isPreviousDate* ||*isDateArea*) && *isVisuallySame* && *isDateArea* && *index-areaIndex==1* && *isPreviousDate* && *isNextDate* && *isDateArea* && *isPreviousDate* && (*dateNum>0* || *dateTypeNum>0*).

For situation (3), the rule is *startWithLi* && (*dateNum>0* || *dateTypeNum>0*) && *wordNum<=DATE_ITEM_WORD_NUM*.

Rule 2: PlaceItem complete rule: PI classification usually has high precision. Some key words and text content features can determine whether a R block is a missed PI block. The rule is: *placeNum>1* && *wordNum<PLACE_WORD_NUM*.

Rule 3: Area complete rule: A AR text block usually has big font size. Therefore, its complete rule is *isHeader=true*.

Rule 4: Topic complete rule: A TO text block has typical context features. It begins with a <li> tag, and if its neighbor block is TO, it is possible TO block too. The complete rule is: (*isTopicArea* || *isPreviousTopic*) && (*isVisuallySame*) && (*isTopicArea*) && (*index-areaIndex=1*) && *startWithLi*.

Rule 5: Position complete rule: A PO text block usually has big font size and few words, and it is also has key words about position. The complete rule is *isTitle* && *wordNum < POSITION_WORD_NUM* && *positionNum > 0*.

Rule 6: PeopleItem complete rule: A PE block should consider its context. Usually, it follows PO blocks. Some PE blocks begin with <li> tag, and some PE blocks contain key words about universities or companies. Most PE blocks have capital letters and short content. The complete rule is: (*isPeopleArea*||*isPreviousPeople*) && *isVisuallySame* && *isPeopleArea* && *index-areaIndex=1* && *startWithLi* && *isPreviousPeople* && *institutionNum>0* && *wordToName < PEOPLE_WORD_TO_ NAME* && *wordNum < PEOPLE_WORD_NUM*.

Rule 7: Paper complete rule: A PA block usually begins with <li>, and its list is similar to TO blocks. The corresponding complete rule is: (*isPaperArea*||*isPreviousPaper*) && *isVisuallySame* && *nameNum> TITLE_UPPER_ WORD_NUM* && *startWithLi* && *nameNum> TITLE_UPPER_WORD_NUM* && *isPaperArea* && *index-areaIndex=1* && *nameNum> TITLE_UPPER_WORD_NUM*.

After the post-processing, initial classification results will be improved greatly. For the reason that classified text blocks describe a special information, these blocks are the academic information we want to extract. Namely, this paper solves an information extraction problem by transforming it into a text block classification problem.

6   Experiments

6.1   System Implementation and Data Set

The system is mainly implemented in Java, and the page segmentation module is implemented in C#. We also use Weka, an open-source machine learning library, to classify text blocks. Our experimental results are obtained on a PC with 2.40 GHz CPU, 2 GB RAM and Windows 7.

Our system can find 50 academic conference websites per hour, and process a conference Web page in average 1.02 s, which contains 0.24 s for segmenting the Web page, 0.60 s for classifying text blocks and 0.18 s for post-processing. Therefore, our system is efficient for handling conference Web pages.

We collect 50 academic conference Web sites in computers science field, which has 283 pages and 10028 labeled text blocks. In order to evaluate our approach, 10 students manually tag all the text blocks as reference results. We randomly select 10 sites which contain 62 pages as training dataset for constructing Bayes Network model. Other 40 conference Websites are used as test dataset. We use Precision, Recall and F1-Measure as criteria to measure the system performance.

6.2   Experiment Results of Crawling Academic Conference Websites

In Figure 5, red nodes are 252 rank A conferences, green nodes are 96 rank B conferences, and blue nodes are 113 rank C conferences. It shows the external link graph between difference rank conferences and internal link graph for each rank conferences. We can observe two facts: (1) There are many links between academic conference websites. (2) High quality conferences such as rank A have few links to other rank conferences, and low quality conferences such as rank C have a lot of links to rank A and rank B conferences. It means that if we want to crawl high quality conferences as much as possible, we should have enough seeds of rank A. (3) From the three internal link graphs, we can see that only few nodes have a lot of links to other nodes. These nodes are the conferences with long history, then new conferences will have links to previous conferences. It means we can easily crawl a same series conferences, but it is difficult to crawl the conferences which are not related.

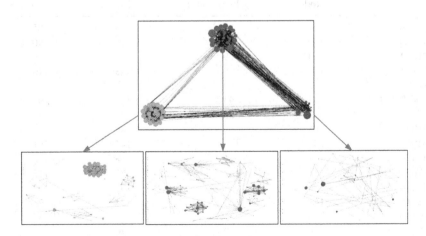

Fig. 5. External and internal link graph of different rank conferences

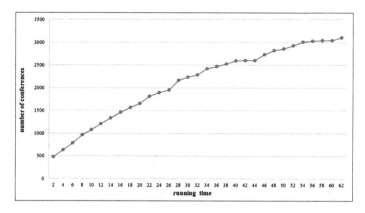

Fig. 6. Crawling performance results

Figure 6 shows that our crawling method can steadily find more than 3109 conference websites during 62 h. It means that we can find 50 new conference websites in one hour.

6.3   Experiment Results of Extracting Academic Information

First experiment is verifying the complete tree. Table 3 shows the vision tree results generated by VIPS and the complete tree results generated by our new algorithm on 15

Table 3. Experimental results of VIPS complete tree

Websites	Vision tree by VIPS		VIPS complete tree	
	Nodes	Leaf nodes	Nodes	Leaf nodes
AAAI-10	106	82	109	85
CIKM-11	109	80	119	90
ICDE-10	94	69	101	76
ICDM-10	131	98	143	110
ICSE-11	145	99	153	107
INFOCOM-10	143	93	143	93
SIGIR-11	133	86	133	86
SIGMOD-10	86	59	86	59
SOSP-11	147	101	147	101
VLDB-10	117	81	123	87
AAAI-11	153	108	157	112
ACL-11	178	137	193	152
AIDE-11	170	121	174	125
ASAP-10	50	31	59	40
ASPLOS-11	60	39	76	55

conference websites. We can see that the complete trees have more leaf nodes than vision trees. It means our algorithm can find more text blocks than VIPS.

Experimental results of removing noise blocks are given in Table 4. We can observe some facts: (1) There are many noise blocks in the complete tree. In some websites, almost half of all blocks are noise blocks. (2) Our removing noise method can remove average 39 % noise nodes and 51 % noise leaf nodes. Therefore, it will reduce the number of nodes should be processed in extraction and improve the efficiency.

The second experiment is the comparison between initial classification results and the results after post-processing. The results are obtained on 20 randomly websites. In Table 5, A1 refers to the initial classification results, and A2 refers to the post-processing results. It can be seen that the classification results are changed after post-processing. Table 6 further evaluate all the results in Table 5. We have two conclusions: (1) The initial classification results only have average 0.75 precision, 0.67 recall and 0.68 F1-measure. After post-processing, the classification results are improved to average 0.96 precision, 0.98 recall and 0.97 F1-measure. Therefore, the post-processing key roles in academic information extraction. (2) Some text blocks like DI, PO, PE and TO, which have clear vision and text content features, have better classification results. The average F1-measure on these blocks is 0.99.

Table 4. Experimental results of removing noise from VIPS complete tree

Websites	Initial complete tree		Remove noise complete tree		Removed node/Before removed	
	Nodes	Leaf nodes	Nodes	Leaf nodes	Nodes	Leaf nodes
AAAI-10	109	85	45	30	0.59	0.65
CIKM-11	119	90	64	32	0.46	0.64
ICDE-10	101	76	46	22	0.54	0.71
ICDM-10	143	110	105	70	0.27	0.36
ICSE-11	153	107	80	37	0.48	0.65
INFOCOM-10	143	93	103	69	0.28	0.26
SIGIR-11	133	86	75	42	0.44	0.51
SIGMOD-10	86	59	49	27	0.43	0.54
SOSP-11	147	101	125	85	0.15	0.16
VLDB-10	123	87	42	17	0.66	0.80
AAAI-11	157	112	102	68	0.35	0.39
ACL-11	193	152	158	110	0.18	0.28
AIDE-11	174	125	108	71	0.38	0.43
ASAP-10	59	40	46	17	0.22	0.58
ASPLOS-11	76	55	39	17	0.49	0.69
Average					**0.39**	**0.51**

Table 5. Experimental results of text blocks classification

Websites	DI		AR		TO		PO		PE		PI		CO	
	A1	A2	A1	A2	A1	A2	A1	A2	A1	A2	A1	A2	A1	A2
AAAI-11	51	58	16	16	107	81	36	16	75	48	0	0	15	11
ASAP-11	5	10	7	7	14	4	8	4	57	58	2	3	5	0
ASPLOS-11	24	43	20	24	35	11	5	3	7	7	7	7	10	7
CCS-11	19	16	9	12	10	0	8	1	4	0	0	0	3	2
CGI-11	9	16	9	10	22	20	5	5	119	122	0	0	10	7
COCOON-11	9	12	6	7	12	12	2	1	28	28	0	0	3	2
FOCS-12	3	3	3	4	1	0	5	4	10	49	0	0	1	1
HT-12	13	14	6	6	67	64	8	7	53	53	0	0	0	2
ICPADS-12	21	24	4	4	22	20	0	0	0	0	0	0	2	1
IFIP-12	12	28	10	10	12	11	14	11	68	67	0	0	2	1
IPDPS-12	15	16	11	14	1	0	10	9	174	173	0	3	6	2
ISCA-12	15	14	4	6	11	10	0	0	7	0	2	4	10	4
LOPSTR-12	11	15	4	4	5	14	3	4	62	68	0	1	2	2
NIPS-12	12	22	3	1	28	10	4	2	65	63	0	0	2	1
PKDD-12	28	29	11	12	18	0	8	7	30	32	0	1	7	7
PODC-12	15	22	6	4	31	30	0	0	0	0	0	1	8	2
CRYPTO-12	9	19	4	5	10	0	9	4	44	98	1	6	1	1
ICPR-12	28	33	6	6	17	24	1	0	16	0	0	0	5	1
SPAA-12	3	8	6	6	2	19	13	8	72	62	0	1	6	3
STOC-12	8	17	5	4	1	17	2	1	15	25	0	1	7	6

Table 6. Statistics of text blocks classification

Category	Initial classification			Post-processing		
	Precision	Recall	F1	Precision	Recall	F1
DI	0.95	0.75	0.84	0.96	0.99	0.98
AR	0.84	0.91	0.87	0.92	0.98	0.95
TO	0.90	0.41	0.57	0.99	0.99	0.99
PO	0.80	0.69	0.74	0.99	0.98	0.99
PE	0.85	0.60	0.71	0.99	0.99	0.99
PI	0.79	0.72	0.75	0.97	0.97	0.97
CO	0.35	0.80	0.49	0.91	0.95	0.93
PA	0.50	0.50	0.50	1.00	1.00	1.00
Average	0.75	0.67	0.68	**0.96**	**0.98**	**0.97**

7 Conclusions

This paper proposes a new method to find and extract useful academic information from conference Web pages automatically. The conference web pages are collected by a lightweight crawling method. Given a sample conference Web page, it is segmented into a set of text blocks using an algorithm which combines vision-based segmentation method and DOM-based segmentation method. Text blocks are classified into predefined categories using bayesian network. Post-processing on the initial classification results can improve the classification. Finally, we integrate extracted information from a conference website to obtain clean and high quality academic data.

References

1. Tang, J., Zhang, J., Yao, L., Li, J., et al.: ArnetMiner: extraction and mining of academic social networks. Presented at the Proceedings of the 14th ACM SIGKDD International Conference on Knowledge Discovery and Data Mining, Las Vegas, Nevada, USA (2008)
2. Chang, C.-H., Kayed, M., Girgis, M.R., Shaalan, K.: A survey of web information extraction systems. IEEE Trans. Knowl. Data Eng. 18, 1411–1428 (2006)
3. Laender, A., Ribeiro-neto, B.A., da Silva, A.S., Teixeira, J.S.: A brief survey of web data extraction tools. SIGMOD Record 31, 84–93 (2002)
4. Kleinberg, J.M.: Authoritative sources in a hyperlinked environment. Presented at the Proceedings of the Ninth Annual ACM-SIAM Symposium on Discrete Algorithms, San Francisco, California, United States (1998)
5. Flake, G.W., Lawrence, S., Lee Giles, C., Coetzee, F.M.: Self-organization and identification of web communities. Computer 35, 66–71 (2002)
6. Cai, D., Yu, S., Wen, J.-R., Ma, W.-Y.: VIPS: a vision-based page segmentation algorithm. Microsoft Technical Report (2003)
7. Liu, W., Meng, X., Meng, W.: ViDE: a vision-based approach for deep web data extraction. IEEE Trans. Knowl. Data Eng. 22, 447–460 (2010)
8. Quinlan, J.R.: C4.5: Programs for Machine Learning. Morgan Kaufmann Publishers Inc., San Francisco (1993)
9. Hastie, T., Tibshirani, R.: Discriminant adaptive nearest neighbor classification. IEEE Trans. Pattern Anal. Mach. Intell. 18, 607–616 (1996)
10. Hand, D.J., Yu, K.: Idiot's Bayes—not so stupid after all? Int. Stat. Rev. 69, 385–398 (2001)

TagRank: A New Tag Recommendation Algorithm and Recommender Enhancement with Data Fusion Techniques

Feichao Ma[1(✉)], Wenqing Wang[1], and Zhihong Deng[2]

[1] Administrative Center of China Academic
Library and Information System (CALIS), Peking University, Beijing, China
{mafc, wangwq}@calis.edu.cn
[2] Key Laboratory of Machine Perception (Ministry of Education),
School of Electronic Engineering and Computer Science, Peking University,
Beijing, China
zhdeng@cis.pku.edu.cn

Abstract. In the era of web2.0, more and more web sites, such as Lastfm, Delicious and Movielens, provide social tagging service to help users annotate their music, urls and movies etc. With the help of tags, users can organize and share their online resources more efficiently. In this paper, we propose a new tag recommendation algorithm TagRank which is based on random walk model. We also explore three data fusion techniques to make more powerful hybrid tag recommenders using TagRank, two collaborative filtering based algorithms and three tag popularity based algorithms. In order to find appropriate individual recommenders to make hybrid, we propose a greedy selection algorithm. We test and verify our proposed TagRank and greedy selection algorithm on three real-world datasets and experimental results show that our methods are efficient in terms of precision, recall and F1.

Keywords: TagRank · Random walk model · Tag recommendation · Data fusion · Social tagging system

1 Introduction

In the era of web2.0, social tagging service is more and more popular, through which users can add keywords or tags to describe their resources such as urls, music, movies, photos, books and so on. On the one hand, users can tag their resources to make it easier to manage their resources. On the other hand, tags can help other users find interesting resources. What's more, tags can also serve as extra metadata to resources, which can contribute to improving information retrieval and recommendation performance. This becomes even more significant, when dealing with multimedia resources such as music, photos, videos etc.

In social tagging systems, users can annotate resources with any tags freely and arbitrarily. It is this freedom that makes tags numerous and quality various. In this sense, tag recommendation becomes important. Because it can help users

S. Zhou and Z. Wu (Eds.): ADMA 2012 Workshops, CCIS 387, pp. 80–91, 2013.
DOI: 10.1007/978-3-642-41629-3_7, © Springer-Verlag Berlin Heidelberg 2013

select good tags in tagging phase, which will give users guidelines to apply high quality tags. Tag recommendation can also help system to show important tags for resources, so that users can explore good resources through those tags. In this paper, we focus on tag recommenders which can be used to assist users during tagging phase and help system in ranking phase when showing tags for resources.

By extending ItemRank [1] algorithm, we propose a novel tag recommendation algorithm TagRank. Our algorithm is totally different from the one described in [2], which is a statistic based method used to rank web pages in terms of tag frequency and the time when the web pages are annotated. Our algorithm is also different from the TagRanker, which is presented in [3]. TagRanker models tag recommendation as a multi-lable classification problem, and solve this problem through an open source library LIBLINEAR [4]. Our proposed TagRank algorithm is based on random walk model. It can help to find pertinent tags for users to annotate their resources. Different recommendation algorithms usually have different advantages and disadvantages. In order to make different algorithms complement with each other, we explore three data fusion techniques: Borda ranking, Reciprocal ranking and CombSum ranking [5] to make a more powerful hybrid tag recommendation algorithm. During the hybrid process, we propose a greedy selection algorithm to determine which individual algorithms to make the final hybrid one.

The rest of this paper is organized as follows. In Sect. 2, we provide a short review of the related work. Then we describe our proposed tag recommendation algorithm TagRank and hybrid strategies in Sect. 3. In Sect. 4, we present experimental datasets, baseline algorithms, experimental methodology and demonstrate the detailed experimental results. Finally, we conclude the paper in Sect. 5.

2   Related Work

There are extensive research efforts dedicated to do tag recommendation in the literature. Existing methods can be generally classified into co-occurrence based method [6], latent factor model [7,8], and random walk based folksonomy model [9,10].

Co-occurrence based methods are simple, but the information they utilize is limited. In order to guarantee recommendation performance, the recommendation results are usually elaborately adjusted by ad-hoc strategies [6]. Latent factor model takes advantage of matrix factorization [7] or expectation maximization algorithm [8] to map tags and users into a low dimension latent factor space. But the computational cost is usually high thus scaling capability is limited. Random walk based folksonomy models [10] embrace users, resources and tags in a uniformed model. If one wants to recommend resources or tags to users, the preference vector should be carefully designed [11].

A lot of previous work in recommendation and information retrieval community [5,12–14] has demonstrated that data fusion technique can enhance individual algorithm's performance. In this paper, we explore three data fusion techniques and propose a greedy algorithm to select appropriate individuals to make a more powerful hybrid algorithm. Our proposed algorithms are simple and scalable, and experimental results on three real-world datasets demonstrate their efficiency in performance.

3   Our Proposed Algorithms

In this section, we first present a novel tag recommendation algorithm TagRank. Then, we delineate three data fusion techniques, and a greedy selection algorithm determining which individuals to select from a candidate algorithm set to make a more powerful hybrid one through data fusion techniques.

3.1   TagRank: A New Tag Recommendation Algorithm

In this section, we propose a novel tag recommendation algorithm based on random walk model [15]. We'll call this algorithm TagRank in our paper, but it is totally different from the TagRank algorithm described in [2], which is a statistic based method used to rank web pages in terms of tag frequency and the time when the web pages being annotated. There're two important data structures to make TagRank work. One is tag-tag Correlation Graph, C, another is tag Bias Vector $\boldsymbol{B}$.

We can define C and $\boldsymbol{B}$ as:

$$C_u(i,j) = |N_u(i) \cap N_u(j)| \tag{1}$$

$$\boldsymbol{B}_u(\text{i}) = |R_u(i)| \tag{2}$$

where $N_u(i)$ is the set of users who have used tag i, $R_u(i)$ is the set of resources user u annotated with tag i. In other words, $C_u(i,j)$ is the number of users who used both tag i and tag j. $\boldsymbol{B}_u(i)$ is relevant to users. For user u, it means the number of times user u has used tag i. To some extend, $\boldsymbol{B}_u$ can represent a user's preference in using tags.

We can also define C and $\boldsymbol{B}$ as:

$$C_r(i,j) = |N_r(i) \cap N_r(j)| \tag{3}$$

$$\boldsymbol{B}_r(\text{i}) = |U_r(i)| \tag{4}$$

where $N_r(i)$ is the set of resources which have been annotated with tag i, $U_r(i)$ is the set of users annotate resource r with tag i. In other words, $C_r(i,j)$ is the number of resources which are annotated by both tag i and tag j. $\boldsymbol{B}_r(i)$ is relevant to resources. For resource r, it means the number of times resource r has been annotated with tag i. Similarly, $\boldsymbol{B}_r$ can suggest a resource's preference being annotated with each tag.

The rationale behind TagRank is that, Bias Vector $\boldsymbol{B}$ represents some preference for each tag and Correlation Graph C describes the correlation strength of two arbitrary tags, we can propagate the preference through Correlation Graph

C in some way to get good tags highly pertinent to the preference expressed in the Bias Vector $\boldsymbol{B}$. This scenario is very similar to Topic Sensitive PageRank [16], so we can formalize the propagation process as:

$$\boldsymbol{TR} = \alpha \cdot \hat{C} \cdot \boldsymbol{TR} + (1-\alpha) \cdot \hat{\boldsymbol{B}} \tag{5}$$

where $\boldsymbol{TR}$ is a vector, in which an entry, $\boldsymbol{TR(i)}$, is the importance score for tag i. α is PageRank const factor and is empirically set to 0.85 [15]. $\hat{C}$ and $\hat{\boldsymbol{B}}$ is constructed by normalizing C_u(or C_r) and $\boldsymbol{B}_u$(or $\hat{\boldsymbol{B}_r}$) as follows:

$$\hat{C}(i,j) = C_u(i,j)/\sum_{j=1}^{|T|} C_u(i,j) \quad or \quad \hat{C}(i,j) = C_r(i,j)/\sum_{j=1}^{|T|} C_r(i,j) \tag{6}$$

$$\hat{\boldsymbol{B}}(\mathrm{i}) = \boldsymbol{B}_u(\mathrm{i})/\sum_{i=1}^{|T|} \boldsymbol{B}_u(i) \quad or \quad \hat{\boldsymbol{B}}(\mathrm{i}) = \boldsymbol{B}_r(\mathrm{i})/\sum_{i=1}^{|T|} \boldsymbol{B}_r(i) \tag{7}$$

For simplicity, We'll call the definition in (1) and (2) U-Rank, the definition in (3) and (4) R-Rank. U-Rank will generate $\boldsymbol{TR}$ for every user. Similarly, R-Rank will generate $\boldsymbol{TR}$ for every resource. Given an user resource pair $< u, r >$, U-Rank will get user u's $\boldsymbol{TR}$ and pick the top N most important tags as recommendation, while R-Rank will get r's $\boldsymbol{TR}$ and pick the top N most important tags as recommendation.

3.2   Hybrid Strategy

Generally speaking, if different recommendation models use different information to do recommendations, applying data fusion technique [5,12] to their recommendation results can bring better performance [13].

In our experiments, we try three data fusion techniques: Borda ranking, Reciprocal ranking and CombSum ranking, because of their simplicity and efficiency.

In tag recommendation scenario, Borda ranking can be formalized as:

$$RS(t_i) = \sum_{j} (\mathrm{N_j} - position(t_{ij})) \tag{8}$$

Reciprocal ranking can be formalized as:

$$RS(t_i) = \sum_{j} (1/position(t_{ij})) \tag{9}$$

CombSum ranking can be formalized as:

$$RS(t_i) = \sum_{j} \frac{score(t_{i,j}) - min(j)}{max(j) - min(j)} \tag{10}$$

The meaning of the symbols in Eqs. (8)–(10): $RS(t_i)$ is the final ranking score for t_i. $position(t_{ij})$ is the position of t_i ranked by the jth recommendation algorithm. N_j is length of recommendation list generated by the jth recommendation algorithm. $score(t_{i,j})$ is the ranking score the jth algorithm assigns to t_i. $min(j)$ is the minimum ranking score the jth algorithm assigns to selected tags, while the $max(j)$ is the maximum one.

Given individual algorithms' recommendation ranking lists, our hybrid strategy is to merge them and using the above equations to re-rank every item in the ranking lists to get the final one. Based on the final ranking list, we can do recommendation in the same way with the individuals.

Algorithm 1. Greedy Selection Algorithm.

Input:

    The set of candidate individual recommenders, C;

    The data confusion technique, ie. Borda ranking, Reciprocal ranking or CombSum ranking, F;

    Evaluation method, E;

    Validation dataset, V;

Output:

    The set of individual recommenders R, which guarantees to improve performance on E by F;

1: Init $R = \emptyset$

2: **while** TRUE **do**

3:    $MaxLift = 0$;

4:    $SelectedIndividual = $ NULL;

5:    **for all** $c \in C - R$ **do**

6:        $H1 = F(R)$; $M1 = E(H1)$;

7:        $D = R \cup \{c\}$;

8:        $H2 = F(D)$; $M2 = E(H2)$;

9:        $Lift = M2 - M1$;

10:       **if** $Lift > MaxLift$ **then**

11:          $MaxLift = Lift$;

12:          $SelectedIndividual = c$;

13:       **end if**

14:    **end for**

15:    **if** $MaxLift > 0$ **then**

16:       $R = R \cup \{SelectedIndividual\}$

17:    **else**

18:       **return** R

19:    **end if**

20: **end while**

Data fusion technique is simple and easy to use, but when there are a lot of individual algorithms, using all of them to make a hybrid recommendation algorithm may not lift and sometimes may hurt performance. In order to determine which individual algorithm to use, we design a greedy selection algorithm.

This algorithm works as follows: It first selects the top performing individual algorithm, and iteratively adding the individual that most improve the currently best performance until there is no improvement. The finally selected individual algorithms will be used to make the final hybrid one. The pseudo code is shown in Algorithm 1.

4   Experiments

In this section, we first describe our experimental setup, including datasets description and our pre-process methods, baselines, and evaluation meatures. Then, we report our experimental method and demonstrate experimental results.

4.1   Experimental Setup

Datasets. Our experiments are conducted using data from three large real-world social annotation systems, which are delicious,[1] lasftm[2] and movielens.[3] They are opened for research use in HetRec 2011 [17]. On all datasets we generate p-cores [9,18]. In this way, some users, resources and tags are removed from the dataset to produce a dense one that guarantees each user, resource and tag occur in at least p annotations. We define an annotation as a user, a resource annotated by this user and all tags this user applied to this resource. By using p-cores, the size of each dataset is dramatically reduced allowing the application of recommendation techniques that would otherwise be computationally impractical, and by removing rarely occurring users, resources and tags, noise in the data can be considerably reduced.

We apply 9-cores to delicious dataset, and 5-cores to lastfm and movielens datasets. In this way, we can get the dense part of each datasets and the size of the resulting datasets is almost in the same scale in terms of the number of users, resources and tags. Statistics of the original dataset and p-core pre-processing one are in Table 1.

Table 1. Dataset statistics

Dataset	p-core	#users	#resources	#tags	#tag annotations
Delicious	-	1867	69223	4089	1867
	9	1779	8427	2243	1642
Lastfm	-	1892	12523	9749	1892
	5	1600	6691	1965	1528
Movielens	-	2113	5908	9079	2113
	5	729	1958	1215	669

---

[1] http://delicious.com/

[2] http://cn.last.fm/

[3] http://movielens.umn.edu/

Baselines. In this section, we give a short review of two kinds of tag recommendation algorithms: One is tag popularity based, another is collaborative filtering based. They are both carefully described in [13]. They'll, on the one hand, serve as baselines to validate the efficiency of our proposed TagRank algorithm, and on the other hand as candidate individuals to make our more powerful hybrid algorithms.

Given an user resource pair $< u, r >$, Tag popularity algorithm recommends the most popular N tags. Resource-Tag popularity algorithm recommends the most popular N tags which are used to annotate r. User-Tag popularity algorithm recommends the most popular N tags which are used by user u. For simplicity, we'll call the Tag popularity, Resource-Tag popularity, User-Tag popularity recommendation methods T-Pop, RT-Pop, UT-Pop respectively.

There are two kinds of collaborative filtering tag recommendation methods. One is user based, another is item based. For simplicity, in User-based collaborative filtering model, when users are modeled in resource space we call it R-UCF; When users are modeled in tag space we call it T-UCF. Similarly, in Item-based collaborative filtering model, when resources are modeled in user space we call it U-ICF; When resources are modeled in tag space we call it T-ICF.

Evaluation Measures. The recommenders are evaluated on their ability to recommend tags for a user resource pair. The user and resource for each annotation were submitted to the recommenders and the recommenders returned a set of tags, T_r. These were then evaluated against the tags in the holdout annotation, T_h.

Precision is one common metric for measuring the usefulness of recommendation algorithms. It measures the percentage of items in the recommendation set that appears in the holdout set. Precision measures the exactness of the recommendation algorithm and is defined as:

$$Precision = |T_r \cap T_h|/|T_r| \tag{11}$$

Recall is another common metric for evaluating the utility of recommendation algorithms. It measures the percentage of items in the holdout set that appear in the recommendation set. Recall is a measure of completeness and is defined as:

$$Recall = |T_r \cap T_h|/|T_h| \tag{12}$$

F1 embraces both precision and recall. It is an overall performance measure on both exactness and completeness of a recommendation algorithm and is defined as:

$$F1 = \frac{2Precision \cdot Recall}{Precision + Recall} \tag{13}$$

Precision, Recall and *F*1 will vary depending on the size of recommendation set. In the following experiments, we get *Precision* and *Recall* with recommendation sets of size one to ten, with which we can plot a precision-recall curve for every algorithm. For each size from one to ten, we can also get the *F*1 metric, and then we average them to get an average $F1$ for every algorithm. Precision-recall curve gives a detailed picture, while $F1$ presents an aggravated view of every algorithm's performance.

4.2 Experimental Methods and Results

In this section we present three experiments to evaluate our algorithms' performance. Experiment one is used to compare our proposed TagRank algorithm with baselines. Experiment two dedicates to show our proposed TagRank can boost other algorithms performance through data fusion techniques. Experiment three validates the greedy selection algorithms' efficiency in selecting good candidates for data fusion techniques. All experiments are carried out on the three p-core preprocessed real-world datasets with five folds cross validation, and the results are averaged over each user, then over the final five folds.

In experiment one, we compare our two TagRank algorithms, R-Rank and U-Rank, with the four collaborative filtering based algorithms and three tag popularity based algorithms, and the results are shown in Fig. 1. In experiment two, we use the three data fusion techniques to hybrid R-Rank and U-Rank with T-ICF. Every data fusion technique makes two hybrid tag recommenders, for example, Borda ranking makes BD-T-ICF+R-Rank and BD-T-ICF+U-Rank. Similarly, Reciprocal ranking makes RP-T-ICF+R-Rank and RP-T-ICF+U-Rank; CombSum ranking makes CS-T-ICF+R-Rank and CS-T-ICF+U-Rank. In experiment three, we use one fold as validation dataset and merge other four folds

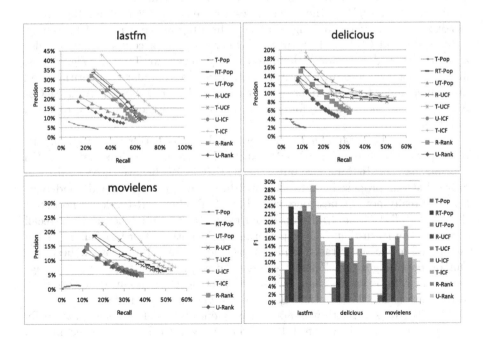

Fig. 1. Experiment one's results

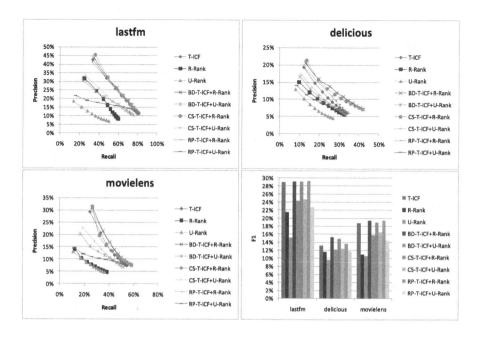

Fig. 2. Experiment two's results

as training dataset to run our greedy selection algorithm. The selecting results are listed in Table 2. With the selecting results, we use Borda ranking, Reciprocal ranking and CombSum ranking to make hybrid tag recommenders BD-Greedy, RP-Greedy, and CS-Greedy respectively. For purpose of comparation, we also apply the three data fusion techniques directly to the two TagRank algorithms, four collaborative filtering algorithms and three tag popularity algorithms, all nine algorithms, to make hybrid tag recommenders BD-All, RP-All and CS-All. Finally, we perform five folds cross validation and the results are reported in Fig. 3.

As the $F1$ bar graph in Fig. 1 shows, R-Rank is better than U-Rank. R-Rank performs better than UT-Pop and T-Pop in all the three datasets. On delicious dataset, R-Rank also outperforms U-ICF. We believe these results can prove that our proposed TagRank algorithm is able to recommend good tags. But it is really not the best one, since both U-Rank and R-Rank are no better than T-ICF, T-UCF and RT-Pop in our experiments. From the precision-recall curve, we can learn that, T-ICF performs significantly better than all others on lastfm and movielens dataset, while on delicious dataset T-UCF is the best one. Our proposed algorithm R-Rank seems good at predicting tags when few tags are recalled, for example when recall level is below 20 %, it can beat U-ICF on lastfm dataset and beat both U-ICF and R-UCF on delicious dataset.

Our TagRank algorithm is not the best one, but as Fig. 2 shows, it really can boost the best algorithm T-ICF significantly with data fusion techniques. R-Rank lifts the $F1$ of T-ICF on lastfm dataset about 0.68 % through Reciprocal

Table 2. Greedy selecting results

Data fusion/dataset	Delicious	Lastfm	Movielens
Borda ranking	T-UCF, T-ICF	T-UCF, T-ICF, U-ICF	T-UCF, T-ICF
Reciprocal ranking	T-UCF	T-UCF, T-ICF, U-ICF	T-UCF, T-ICF
CombSum ranking	RT-Pop, T-UCF	T-UCF, T-ICF	T-UCF, T-ICF, U-ICF

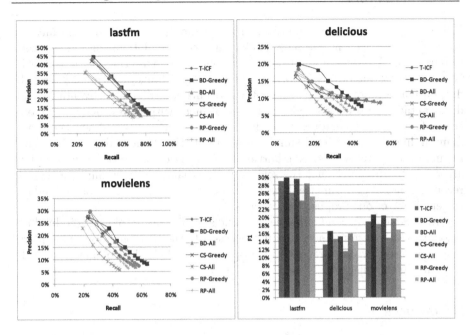

Fig. 3. Experiment three's results

ranking, and increase the $F1$ of T-ICF on delicious and movielens datasets about 2.02 % and 3.52 % respectively through Borda ranking. U-Rank doesn't bring benefit and even hurt the performance of T-ICF in terms of $F1$. The precision-recall curves tell us, R-Rank can lift the precision of T-ICF on almost every recall level on all of the three datasets. From the experiment two's results, we can also learn that, different datasets may fit different data fusion techniques, and the position based technique, Reciprocal ranking and Borda ranking, seems more appropriate to make hybrid of TagRank and collaborative filtering algorithms, which we attribute to the difference of their ranking score's meaning.

As the $F1$ bar graph in Fig. 3 shows, simply feed all nine algorithms to data fusion algorithms hurt recommendation results at most time, while using greedy selection algorithm can persistently improve recommendation performance. From the precision-recall curve, we can learn that, the hybrid algorithms resulting from our greedy selection algorithm can significantly improve precision level on large recall level, while may brings little hurt in precision when there is only few tags recalled. When using data fusion technique to improve performance, efforts should be spared to select good data fusion techniques. For example, on delicious dataset, we may select BD-Greedy because it lifts precision on all recall level.

5 Conclusions

In this paper we propose a novel tag recommendation algorithm TagRank. It aims to provide relevant and high quality tag recommendations for a given user-resource pair. Although it can't beat some collaborative filtering methods, but it is able to boost the best collaborative filtering algorithm's performance through data fusion techniques. Since different algorithms have different advantages and disadvantages, in order to make them complement with each other, we explore three data fusion techniques to make hybrid recommenders. When there are many individual algorithms, simply using all of them to feed data fusion algorithm may not improve and even hurt performance, we propose a greedy selection algorithm to guarantee that the selected individuals can make improvement in recommendation performance. Experiments on three real-world datasets validate the efficiency of our proposed algorithms.

Although data fusion technique is simple and efficient in robusting individual recommender's performance, it is not necessarily the best one. We will explore more advanced hybrid strategies to propose better hybrid recommendation algorithms in the future.

References

1. Gori, M., Pucci, A.: Itemrank: a random-walk based scoring algorithm for recommender engines. In: International Joint Conference on Artificial Intelligence, pp. 2766–2771 (2007)
2. Jie, S., Chen, C., Hui, Z., Rong-Shuang, S., Yan, Z., Kun, H.: Tagrank: a new rank algorithm for webpage based on social web. In: Proceedings of the 2008 International Conference on Computer Science and Information Technology, pp. 254–258 (2008)
3. Montanes, E., Quevedo, J.R., Daz, I., Cortina, R., Alonso, P., Ranilla, J.: Tagranker: learning to recommend ranked tags. Log. J. IGPL / Bull. IGPL **19**, 395–404 (2011)
4. Fan, R.E., Chang, K.W., Hsieh, C.J., Wang, X.R., Lin, C.J.: Liblinear: a library for large linear classification. J. Mach. Learn. Res. **9**, 1871–1874 (2008)
5. Shaw, J.A., Fox, E.A.: Combination of multiple searches. In: Text REtrieval Conference (1994)
6. Sigurbjörnsson, B., van Zwol, R.: Flickr tag recommendation based on collective knowledge. In: Proceedings of the 17th International Conference on World Wide Web, WWW '08, pp. 327–336. ACM, New York (2008)
7. Symeonidis, P., Nanopoulos, A., Manolopoulos, Y.: Tag recommendations based on tensor dimensionality reduction. In: Conference on Recommender Systems, pp. 43–50 (2008)

8. Wetzker, R., Umbrath, W., Said, A.: A hybrid approach to item recommendation in folksonomies. In: Proceedings of the WSDM '09 Workshop on Exploiting Semantic Annotations in Information Retrieval. ESAIR '09, pp. 25–29. ACM, New York (2009)

9. Jäschke, R., Marinho, L., Hotho, A., Schmidt-Thieme, L., Stumme, G.: Tag recommendations in folksonomies. In: Kok, J.N., Koronacki, J., Lopez de Mantaras, R., Matwin, S., Mladenič, D., Skowron, A. (eds.) PKDD 2007. LNCS (LNAI), vol. 4702, pp. 506–514. Springer, Heidelberg (2007)

10. Hotho, A., Jäschke, R., Schmitz, C., Stumme, G.: Information retrieval in folksonomies: search and ranking. In: Domingue, J., Sure, Y. (eds.) ESWC 2006. LNCS (LNAI), vol. 4011, pp. 411–426. Springer, Heidelberg (2006)

11. Kim, H.N., El Saddik, A.: Personalized pagerank vectors for tag recommendations: inside folkrank. In: Proceedings of the Fifth ACM Conference on Recommender Systems, RecSys '11, pp. 45–52. ACM, New York (2011)

12. Nuray, R., Can, F.: Automatic ranking of information retrieval systems using data fusion. Inf. Process. Manag. **42**, 595–614 (2006)

13. Gemmell, J., Schimoler, T., Mobasher, B., Burke, R.D.: Hybrid tag recommendation for social annotation systems. In: International Conference on Information and Knowledge Management, pp. 829–838 (2010)

14. Pazzani, M.J.: A framework for collaborative, content-based and demographic filtering. Artif. Intell. Rev. **13**, 393–408 (1999)

15. Page, L., Brin, S., Motwani, R., Winograd, T.: The pagerank citation ranking: bringing order to the web. Technical Report 1999-66, Stanford InfoLab (1999)

16. Haveliwala, T.H.: Topic-sensitive pagerank. In: World Wide Web Conference Series, pp. 517–526 (2002)

17. Cantador, I., Brusilovsky, P., Kuflik, T.: 2nd workshop on information heterogeneity and fusion in recommender systems (hetrec 2011). In: Proceedings of the 5th ACM conference on Recommender systems. RecSys 2011. ACM, New York (2011)

18. Batagelj, V., Zaversnik, M.: Generalized cores. Computing Research Repository (2002)

Social Information Diffusion

Multi-Influences Co-existence Based Independent Cascade Model and Conflict Resolution Strategy in Social Networks

Jing Wanjing[⊠], Chen Hong, and Zhao Linlu

Renmin University of China, Beijing, China
jingwanjing@126.com, chong@ruc.edu.cn, super_zhao@126.com

Abstract. Social Networks Influence Maximization Problem has been studied extensively and most have focused on the single influence, while few studies have focused on the Multi-influences co-existence based influence diffusion. In this paper, we extend the traditional Independent Cascade Model (*ICM*) and propose the Multi-influences based Independent Cascade Model (*MICM*), put forward two kinds of novel conflict strategy algorithms, which are the Largest Neighbor Conflict Strategy (*LNCS*) and Conflict Vector Transform Strategy (*CVTS*).We also elaborate conflict strategies from three different perspectives of propagation rules, customers and producers. To illustrate these issues, we conduct experiments with data from four real datasets, evaluate the performances of the proposed model and algorithms, demonstrate that the final numbers of influential nodes are progressive when *MICM* uses three different conflict selection strategies.

Keywords: Social networks · Multi-influences · *MICM* · Max degree algorithm · Conflict strategy

1 Introduction

Social Network is a network system which consists of social relations between individual members. In recent years, along with the Internet and Web 2.0's booming development, with the social network and social media website's constantly emerging, which provide a good platform for people's communication, knowledge sharing, and information diffusion, produce a significant impact on people's daily life and behavior. Social Network Influence is the behavior change influenced by other individual members in social network.

Social Networks Influence Maximization Problem has been widely studied, here we borrow the growth function $F(k,G,M,\Gamma)$ [6] to describe the problem, among which k is the initial number of active nodes, G is the network structure, M is the based propagation model, and Γ is the adopted influence maximization algorithm for selecting the initial active nodes. Therefore, influence maximization problem can be expressed as how to effectively choose the four parameters of growth function so that the final active nodes can reach scale maximization.

S. Zhou and Z. Wu (Eds.): ADMA 2012 Workshops, CCIS 387, pp. 95–105, 2013.
DOI: 10.1007/978-3-642-41629-3_8, © Springer-Verlag Berlin Heidelberg 2013

At present, the most basic and widely studied propagation models are the Linear Threshold Model (*LTM*) [1] and Independent Cascade Model (*ICM*) [1] proposed by Kempe et al. Ji Jinchao et al. [3] present a Complete Cascade Model, in which the influence probability among nodes may change due to the dynamic change of the interaction' intensity between nodes. However, the above models are only for single influence without consideration of multi-influences in social networks. *IBM* problem [4] is about how to choose a positive influence to resist the harm produced by a negative influence. Wang et al. [5] considers the case of simultaneous propagation of two influences in small-world networks. Our studies are different from the above literature researches, and we focus on the multi-influences co-existence based situation where the network has already existed other similar influences when you want to introduce a new influence, we also propose the conflict resolution strategies corresponding with the model.

Specifically speaking, in this paper, we extend the Independent Cascade (*IC*) model [1] and build Multi-influences co-existence based Independent Cascade Model (*MICM*). The propagation mechanism is running like this, when an influence labeled *I* begins to spread, there are already a variety of other *Non-I* similar kind of influences. And it is likely that when influence *I* spreads to a node with another similar kind of influence, which may face an inevitable conflict. Take marketing products as an example of influence propagation, we propose conflict resolution strategies respectively from the propagation rule itself, customers and producers.

In summary, our contributions are threefold. First, we propose a new novel propagation model *MICM* for multi-influences pre-existence; the model extends the conventional *IC* model. Second, we put forward the two algorithms *LNCS* (Largest Neighbor Conflict Strategy) and *CVTS* (Conflict Vector Transform Strategy); the conflict resolution strategy can be expressed in three different perspectives. Third, we conduct experiments in four different orders of magnitude scale dataset, and give comparative analysis on the results of three multi-influences conflict resolution strategies; experimental results show that our model and algorithm have feasibility, effectiveness and progressiveness.

The remainders of this paper are organized as follows. Section 2 reviews related work and backgrounds. Section 3 presents *MICM* model and corresponding conflict selection strategy algorithms. We show the experiments and evaluation in Sect. 4. Finally, we offer conclusions and research directions in Sect. 5.

2 Backgrounds

2.1 Influence Maximization Algorithm

Max-Degree Influence Maximization Algorithm. A social network can be modeled as a graph $G = \{V, E\}$, where V represents the set of nodes, E represents the set of edges. Due to the characteristics of the social network structure, some nodes are with higher influences than other nodes and the nodes with higher degree usually have greater influences. *Max-Degree* Influence Maximization Algorithm has been proposed based on node degree, which has the advantage of low time complexity and could be applied to any large network. The disadvantage of the algorithm is that propagation effect is unstable and problem solution would vary with the change of the network structure.

Climbing Greedy Influence Maximization Algorithm. Kemple and Kleinberg [1] propose a Climbing Greedy Algorithm, as to an initial active set A, if $\sigma(A)$ presents the expectation of active influence nodes at the end of the propagation process, then in each step they choose a maximum marginal gain node $\sigma(A \cup \{v\}) - \sigma(A)$ until having chosen k nodes. The advantage of climbing greedy algorithm is getting a stable solution, which could reach an approximate optimal value within 1-$1/e$ factor. However, the time complexity is very high, for a network of hundreds of nodes, completing the work of searching for the initial node would cost a long time, not to mention hundreds of thousands of large networks. Therefore, the greedy algorithm is only suitable for small scale networks.

Community-Based Influence Maximization Algorithm. Social network has gradually become an important route of transmission due to its broadcasting and sharing. At present, there are two understandings for graph G, one is global graph structure, another is local structure formed by a community-based network architecture. Community-based Influence Maximization Algorithm first uses the existing community detection algorithm to divide the network into several communities, and then applies the influence maximization algorithm in each separate community. Both *OASNET* [6] and *CGA* [2] are algorithms which have adopted community discovery algorithm to solve the problem of influence maximization.

2.2 Related Work

Linear Threshold Model [1]. *LTM* (Linear Threshold Model) represents the difficulty of a node been activated with a threshold θ_v ($\in[0,1]$). The smaller of a node threshold, accordingly the node is easier to be affected by other nodes and vice versa. Suppose a node v has an influence weight $\sum b_{v,w} \geq 0$ from each of its neighbor w, subject to the constraint that $\sum b_{v,w} \leq 1$. For all the neighbors of v, as long as $\sum b_{v,w} \geq \theta_v$ is established, v will become active from a non-active state; the activated v can influence its neighbors in subsequent propagation; the above process will be repeated until no new nodes can be activated.

Independent Cascade Model [1]. *ICM* (Independent Cascade Model) represents the influence from a node v to another node w with a probability $p(v,w)$ ($\in[0,1]$). The greater the probability is, the corresponding node is more likely to be activated successfully. At a time t, suppose that node v is active, if v activates w successfully, then w will become an active node at time $t + 1$. But no matter node v is successful or not, v will no longer try to activate w in subsequent stages. The process continues until there are no new nodes been activated.

Comprehensive Cascade Model [3]. Kempe et al. [10] propose a descending cascade model considering the influence attenuation between nodes. In fact, the influence between nodes not only exists decay, but also can be strengthened or remain unchanged. Based on this idea, [3] puts forward a Comprehensive Cascade Model, which can be expressed in a formula $p_w(v, S) = p_w(v) - k \times |S|/|V|p_w(v)$, where S represents the neighbor set of w, which have tried to activate w but unsuccessfully, $p_w(v, S)$ means the influence probability from an active node v to an inactive neighbor node w, k indicates a random value selected from $\{-1,0,1\}$, $|V|$ is the number of nodes in the real network, $p_w(v)$ is the initial influence probability from node v to node w.

3   MICM Influence Model and Conflict Strategy

In this section, we propose the Multi-influences Independent Cascade Model (*MICM*), at the same time, we put forward two kinds of conflict strategies applied in the influence propagation. One is from the consumer's point of view and named as *LNCS* (Largest Neighbor Conflict Strategy), another is from the producer's perspective and named as *CVTS* (Conflict Vector Transform Strategy). Section 3.1 will give the algorithm pseudo-code. Next, *MICM* will be systematically explained in Sect. 3.2.

3.1   Conflict Strategy Selection Algorithm

LNCS (Largest Neighbor Conflict Strategy)

```
Input: graph G, initial k influence set A, conflict node x.
Output: current influence set T.
1、 T←A, countI←0, countO←0;
2、 FOR Node n : x.adjLink  //visit adjacency list of conflict node x
3、    IF the neighbor of x has the influence labeled I
4、        countI++ ;  //count the number of neighbor nodes labeled I
5、    ELSE IF the neighbor of x has other similar kind of influences
presented as O
6、        countO++ ;  //count the number of neighbor nodes labeled O
7、 ENDFOR
8、 IF countI>countO
9、    relabel the influence of node x as I, and T←T∪x ;
10、RETURN T ;
```

CVTS (Conflict Vector Transform Strategy)

```
Input: graph G, initial k influence set A.
Output: current influence set T.
  1、 T←A, conflictVectorList ← ∅
  2、 FOR Node x : T    //traverse current influence set T
  3、    FOR Node y : x.adjLink    //visit adjacency list of node x
  4、      IF y already has influence labeled 0 in influence propagation
process
  5、        conflictNodeList ← conflictNodeList ∪ y; //save conflict
node
  6、    ENDFOR
  7、    Define conflictVector[] ← {0} , corresponding with conflict
node list
  8、    FOR i=0 TO conflictVector.length
  9、      IF i==0    //conflict vectors all are 0, no need to permutation
 10、       begin simulation propagation;
 11、      ELSE
 12、        FOR j=1 TO i
 13、          conflictVector[j-1] ← 1; //gradually convert 0 to 1
 14、        ENDFOR
 15、        IF i== conflictVector.length    //conflict vectors all are 1,
no need to permutation
 16、          begin simulation propagation;
 17、        ELSE
 18、          getAllOrder(conflictVector,0,conflictVector.length);
//permutation function
 19、          every kind of permutation result begins simulation
propagation;
 20、    ENDFOR
 21、    save the bestConflictVector in simulation propagation;
 22、    FOR i=0 TO bestConflictVector.length
 23、      IF bestConflictVector[i]==1
 24、        label the influence of node y as I, and T←T∪y;
 25、    ENDFOR
 26、 ENDFOR
 27、 RETURN T ;
```

3.2   The Proposal of MICM

The previous social networks influence maximization problems are mostly based on the single influence propagation [10, 11, 13], or based on the mutual resistance of two kinds of symbol influences [4], or based on the simultaneous spread of two kinds of influences [5]. In the paper, we present a new type of multi-influences co-existence based independent cascade propagation model (*MICM*). When a new type of influence

Table 1. The comparison of ICM based and MICM based influence propagation

Influence propagation steps	ICM based influence model	MICM based influence model
Step 1	Build social network graph G'	Build social network graph G (*same*)
Step 2	Select initial k influence set A'	Select initial k influence set A plus conflict strategy S (*different*)
Step 3	Traditional influence propagation model M'	MICM influence propagation model M plus conflict strategy S (*different*)
Step 4	Influence spread and convergency	Influence spread and convergency (*different*)

begins to spread, there have been many other similar influences in the network. At this time, if the influence diffusion is based on the traditional *IC* (Independent Cascade) model, which may generate inevitable conflict. We can apply the strategy algorithms given in Sect. 3.1 to deal with the conflict.

The MICM Based Influence Propagation Process. First let us comb the influence propagation process in social networks. Step one, we should select a reasonable social network graph G, which can be a global graph, or the community based local network structure. The second step, selecting top k most influential nodes as the initial set A, which needs to adopt one kind of influence maximization algorithm (Sect. 2.1). Step three, the spread of initial influence is based on the propagation model mechanism (Sect. 2.2). The last is the influence spread and convergency, the time of convergency is decided by both the influence diffusion mechanism and the number of initial influence nodes.

Similarly, the *MICM* based influence diffusion process also can be recognized as four steps, except a little difference (*shown in* Table 1). The first step is same. Step two, when we choose the top k most influential nodes, which may produce conflicts, we could consider the conflict strategies (*mentioned in* Sect. 3.1) to solve them. The third step, in the process of influence propagation, it would encounter with nodes labeled with other similar influences, for which we could also adopt the conflict strategies. The fourth step involves influence diffusion model mechanism, so it's differ from the former.

Next, we will discuss the conflict resolution strategy from three different aspects, including the propagation rule itself, consumers and producers.

Conflict Resolution Strategy. If we take selling product I for example, assume that there may have other products other than I, such as J, etc. Now we consider the product I's propagation. If somebody is selling the product I, we view it as a producer, and we view the recipient of the product I as a consumer, we regard selling product from the producer to consumer (having accepted product J, or not accepted any product I or J), the spread process from a consumer having accepted product I to other consumers having not accepted any product I or J as the propagation rules themselves.

First of all, in terms of the propagation rule itself, in the influence I's propagation process, if a node labeled with no other influences is reached, then it can be activated. However, if meeting with a node labeled with influence as similar as I, then the propagation will skip it, directly go to the next round of activation.

Secondly, from the perspective of consumers, for general public, they will ask for advices from their neighbors when they first contact a product I. Therefore, if we take this into consideration, in the propagation process of influence I, if a node labeled with no other influences is reached, then it can be activated. Otherwise, if meeting with a node labeled with influences akin to I, then the propagation will need to adopt the *LNCS* (Largest Neighbor Conflict Strategy) given in Sect. 3.1.

Thirdly, from the producers' view, as the sellers, they naturally hope that more and more products can be sold. Therefore, they mainly focus on the final influences scale maximization. Equally, in the propagation process of influence I, if a node labeled with no other influences is reached, then it can be activated. However, if encountering with a node labeled with influence as similar as I, then we could use the *CVTS* (Conflict Vector Transform Strategy) presented in Sect. 3.2.

To sum up, the theoretical contributions of this paper are as follows. First, we offer the algorithm pseudo codes of *LNCS* and *CVTS*. Secondly, we put forward the *MICM* multi-influence model, and make a comparison between the traditional *ICM* and *MICM*; Moreover, we further explain the conflict resolution strategy from three different perspectives.

In the following, we will verify the feasibility and effectiveness of the proposed model and algorithm by conducting experiments and evaluations.

4 Experiments

4.1 Experiment Datasets

We perform experiments on four different orders of magnitude real datasets, and the detailed statistical information is shown in Table 2.

The *Football* dataset is the relationship data of American high school football summer class *A* regular season in 2000, which contains 35 nodes and 118 edges. Each node represents a team, and the edge signifies the link between teams. The *Graph Products* data is about the 2-mode network structure composed of papers and authors, which contains 674 nodes and 314 edges. Papers and authors form the nodes, and their relationships form the edges. The *Geom* dataset is a computational geometry

Table 2. The statistical information of datasets

No.	Datasets	Nodes	Edges
1	Football	35	118
2	Graph products	674	314
3	Geom	7343	11898
4	PGP	10680	24340

cooperation network, which studies the cooperation of authors in this field, consists of 7343 node and 11898 edges. The author represents the node, there is a edge between two authors representing that they have cooperation work. The dataset of *PGP* comes from the user network, which employs the Pretty-Good-Privacy algorithm for exchanging safety information, including 10680 nodes and 24340 edges. Users represent nodes, the secure information exchanges form the edges.

4.2   Experiment Design

The experiments adopt the *Max-Degree* Influence Maximization algorithm, are based on the *MICM* propagation model, and are assisted with the two kinds of conflict resolution strategies given in Sect. 3.1. In the following part, *S1* represents the conflict resolution strategy based on propagation rule itself, *S2* represents the *LNCS* based on consumer, and *S3* represents the *CVTS* based on producer. In the *MICM* based influence diffusion process, we will unify the propagation factor as $p = 0.1$.

The experiments use Java programming language, and based on the platform of *Windows Xp + Eclipse + Jdk 5.0*.

4.3   Experiment Results and Analysis

Figures 1, 2, 3, and 4 show experimental results of dataset 1, 2, 3, 4 respectively. The *x* axis represents the initial *k* influence nodes, the *y* axis represents the final activated nodes at the end of the diffusion. The three colored curves show the different outcomes of adopting conflict strategy *S1, S2, S3*. From the figures we can directly see that the result of *S3*, i.e. using the *CVTS*, is the best of all, *S2* is the second best, and *S1* is the worst.

As for the performance of the above results, it is consistent with the theoretical assumption. For *S1*, which uses the propagation rule itself, as long as meeting with the node labeled with influences akin to *I*, the propagation skips it. However, for *S2*, using the *LNCS*, if encountering with a node labeled with influences as similar as *I*, the node tends to choose a influence according to most of its neighbor's choice, therefore it may

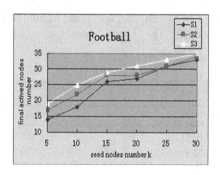

Fig. 1. Dataset 1 influence propagation comparison of different conflict strategies

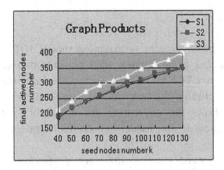

Fig. 2. Dataset 2 influence propagation comparison of different conflict strategies

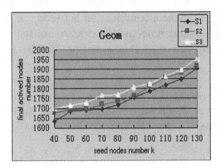

Fig. 3. Dataset 3 influence propagation comparison of different conflict strategies

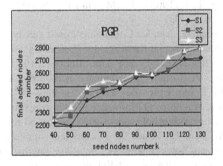

Fig. 4. Dataset 4 influence propagation comparison of different conflict strategies

produce a better result. As to *S3*, which uses the *CVTS*, it makes a conflict list for all the nodes meeting the activation condition, simulates diffusion for every kind of conflict vector transform, finally chooses conflict nodes with the simulative maximized diffusion scale, this strategy is like an exhaustive conflict strategy and could generate the best effectiveness. In the practical application, it's necessary to select suitable conflict strategy to solve the conflict problem in the influence diffusion according to different situations.

5 Conclusion

This article proposes a novel type of propagation model *MICM*, which is under the premise of multi-influences pre-existence, also presents two kinds of conflict resolution strategy algorithms *LNCS* and *CVTS*. further explains the problem from three different perspectives including the propagation rules itself, consumer and producer. The proposed model shows the innovation and scalability. At last, we verify the feasibility and effectiveness of the model and conflict strategies by the experiments and evaluations.

Although we put forward a novel influence propagation model under multi-influence, and the theoretical guess is consistent with the experimental results, there are still many aspects needing to be improved and further explored. For example, the research on the mute-influences pre-existence based influence maximization algorithms; the study on how to make a combination between our work in this paper with community discovery and heterogeneous network; As for the current hot topic of big data, is it possible to dig out useful information for the influence maximization researches.

Acknowledgement. This research was supported by the "Domestic Database High Performance and High Security Key Technology Research" of *HGJ* Important National Science & Technology Specific Projects of China (2010ZX01042-001-002-002).

References

1. Kempe, D., Kleinberg, J.M., Tardos, E.: Maximizing the spread of influence through a social network. In: KDD (2003)
2. Wang, Y., Cong, G., Song, G., Xie, K.: Community-based greedy algorithm for mining top-k influential nodes in mobile social networks. In: Proceedings of the 16th ACM SIGKDD Conference on Knowledge Discovery and Data Mining (2010)
3. Ji, J., Han, X., Wang, Z.: Community influence maximizing based on comprehensive cascade diffuse model. J. Jilin Univ. (Sci. Ed.) **47**(5), 1032–1035 (2009)
4. He, X., Song, G., Chen, W., Jiang, Q.: Influence blocking maximization in social networks under the competitive linear threshold model. arXiv technical report, 21 Oct 2011
5. Wang, N., Wang, X., Sun, Q., Zhao, L.: A computer simulation model of competing products diffusion on small-world network. In: 2008 International Seminar on Future Information Technology and Management Engineering (2008)
6. Cao, T., Wu, X., Wang, S., Hu, X.: OASNET: an optimal allocation approach to influence maximization in modular social network. In: SAC'10, 22−26 March 2010
7. Chen, W., Wang, Y., Yang, S.: Efficient influence maximization in social networks. In: KDD (2009)
8. Lappas, T., Terzi, E., Gunopulos, D., Mannila, H.: Finding effectors in social networks. In: KDD (2010)
9. Li, C.-T., Lin, S.-D., Shan, M.-K.: Finding influential mediators in social networks. In: WWW (2011)

10. Kempe, D., Kleinberg, J., Tardos, É.: Influential nodes in a diffusion model for social networks. In: Caires, L., Italiano, G.F., Monteiro, L., Palamidessi, C., Yung, M. (eds.) ICALP 2005. LNCS, vol. 3580, pp. 1127–1138. Springer, Heidelberg (2005)
11. Kimura, M., Saito, K., Nakano, R., Motoda, H.: Extracting influential nodes on a social network for information diffusion. Data Min. Knowl. Disc. **20**(1), 70–97 (2010)
12. Bharathi, S., Kempe, D., Salek, M.: Competitive influence maximization in social networks. In: Deng, X., Graham, F.C. (eds.) WINE 2007. LNCS, vol. 4858, pp. 306–311. Springer, Heidelberg (2007)
13. Even-Dar, E., Shapira, A.: A note on maximizing the spread of influence in social networks. In: Deng, X., Graham, F.C. (eds.) WINE 2007. LNCS, vol. 4858, pp. 281–286. Springer, Heidelberg (2007)

Herd Behavior Analysis Based on Evolutionary Game Theory in the Decision-Making of Innovation Adoption

Mao Yan-ni[1,2(✉)] and Zhou Lei[1]

[1]School of Information Management, Wuhan University, Wuhan,
People's Republic of China
13813966398@139.com
[2]School of International Economics and Business,
Nanjing University of Finance and Economics, Nanjing,
People's Republic of China

Abstract. Information Cascades are fairly important to individual decision making, however, it of lead to adoption of inferior products. Many researches on herd behavior have limited their scope on the phenomenon in capital markets and the behavior of investors, but few of them focus on the herd behavior of agents. This paper aims to employ the evolutionary game theory to explain the cause of the herd behavior of agents in their decision-making process for innovation adoption. Furthermore, the effects of herd behavior on innovation adoption and diffusion are analyzed. A kind of remuneration incentives mechanism is then presented in order to circumvent the herd behavior of agents. Last but not the least, a case study on the innovation adoption of e-business enterprises in China has been conducted to demonstrate the effectiveness of the proposed methods.

Keywords: Innovation adoption · Herd behavior · Evolutionary game · Incentive mechanism

1 Introduction to Herd Behavior

1.1 Key Concepts of Herd Behavior

Herd behavior usually is related to individuals who ignore their own private information and imitate others behavior because of external interference in the incomplete information environment [1]. According to the previous research, herd behavior is defined as follows: in a condition under which a subsequent actor could observe predecessors, he will make a decision regardless of his personal information during the decision-making process. In a word, when herd behavior appears, individuals tend to act in concert.

S. Zhou and Z. Wu (Eds.): ADMA 2012 Workshops, CCIS 387, pp. 106–115, 2013.
DOI: 10.1007/978-3-642-41629-3_9, © Springer-Verlag Berlin Heidelberg 2013

1.2   Cause of Herd Behavior

Information cascade is an important cause of herd behavior when making a decision related to innovation adoption. Information cascade refers to a rational individual who ignores his personal information when adopting an innovation as an imitation of predecessor's behavior, although he should have taken a completely different behavior according to information obtained. Information cascade is a widespread phenomenon in the financial investment and new technology adoption [2]. Due to the asymmetry of information, the information cascade will affect competition and diffusion of an innovation. Herd behavior is a behavior pattern caused by information cascade. Information cascade produces a herd behavior when individuals choose to adopt a similar behavior in accordance with the behavior of other actors. There are several reasons leading to herd behavior caused by information cascade, among them, they are switching costs and information externalities.

The switching cost is the transition cost paid by consumers who are in the purchase of a commodity to replace another. High switching cost is the characteristic of a lot of technical innovation. Transfer cost at some occasions may be higher than the product price, including the cost of learning, addition purchases of compatible products, cost to create new documents and the cost of network externalities.

High switching costs brings about a high risk to adopting an innovation, because it is difficult to change a decision made. In order to avoid be locked by an inferior technology, wait-and-see strategy is exerted by a potential user as observing forerunner's behavior will expose to more information of innovative technologies to reduce the risk of the decision-making.

The information externality refers to obtaining forerunner's personal information by observing other's action. Decision-maker updates his relative information from information overflow which benefits him finally.

Information externality is a direct cause of the information cascade. Information externality is an important means of communication in the innovation diffusion process; most users ignore their personal information and imitate prior user's decision. As a result, users adopt the same technology in a relatively short period of time.

Information cascade may lead to the contradiction between individual rationality and collective irrationality. The information cascade accompanying innovative technology adopting is caused by incomplete information and asymmetric information. Compared to independent decision-making with the full use of personal information, a decision affected by information cascade is with high risk to decision-making. With the formation of the information cascade, sub-optimal technology may become the mainstream of the market as well as lead to the repeated construction of innovative technology, and only to far away from maximal social welfare.

1.3   Effects of Herd Behavior

Herd behavior is a very common phenomenon in the market of innovative products with more negative impacts on the market instead of positive impacts [3].

First, herd behavior adopter often abandons his own private information to follow others, which may interrupt market information chain as well as destroy market stability.

Second, if the herd behavior exceeds a certain limit, market bubbles and repeated construction are created.

Third, an important basis of the herding behavior is incomplete information. Therefore, even the market information state changes, such as the arrival of new information, herd behavior in general will collapse [4]. It means that instability and vulnerability are characteristics of the herd behavior, which produces the instability and vulnerability of the market directly.

2 Herd Behavior of Agent's Decision-making of Innovation Adoption

2.1 Effects of Herd Behavior of Innovation Adoption and Diffusion

Innovation has an important role in Socio-economic development and industrial competitiveness. The value of innovations appears only when innovations are adopted and accepted by the market. With the separation of corporate ownership and management, decision-making power of business operations to a greater extent transfers to the hands of the agents who become one of the main driving forces of technological innovation and institutional innovation. As Schumpeter pointed out, the entrepreneurs are the advocates and practitioners of innovation activities. The functions of an entrepreneur is through the use of a new invention, or more generally speaking, the use of a new methods in which new products or old have never been produced, or creating new source of raw materials or market of channels of sales, or initiating an evolution to reconstruct the industry.

Firm agents and principals vary in their objectives. Agents tend to follow other agents' investment decisions to avoid risks as well as maximize their own utility, which gives a birth to herd behaviors. Because agent's compensation is performance-related, combined with a desire to keep professional reputation based on his long-term investment revenues, agents generally lack adventurous spirit to follow peer's decision [5, 6]. In order to avoid the investment with high risk which generates adverse effects, agents often choose investment projects with lower level of risk, or delay the confirmation of uncertain projects, or to follow the investment decisions of others. The company loses many innovation and development opportunities because of agent's ostentatious attitude, thereby reducing the potential benefits of the shareholders. Especially under the case of technological reforms or market environment changing rapidly, the negative effects is particularly prominent.

2.2 Herd Behavior of Agents

There are two main causes to contribute to agents' herd behavior in the innovation adoption.

(1) Concern on reputation risk. Scharfstein and Stein proposed a learning model to explain agents' herd behavior. The agents ignore private information and imitate other's decision in order to avoid being deemed incompetent as making a decision different from other managers. In order to protect their own reputation as well as share the blame effect when a wrong investment has been made, agents follow others' opinion to make the same decision. Devenow and Welch pointed out those agents in order to avoid the performance lagging behind peers tend to make similar investment decisions as others.

(2) Compensation mechanism based on relative performance.Maug and Naik hold an opinion that if an agent's compensation depends on his performance relative with other agents, agent incentive mechanism will distort and invalid portfolio may appear, finally herd behavior are in general. Because there is a relationship between the agent and principal, phenomenon as moral hazard and adverse selection exist inevitably. Due to the moral hazard and adverse selection, a contract related to agent's performance is optimal to the principals. On the one hand, such a contract motivates agents to gather information to reduce the risk of moral hazard [7]. On the other hand, excellent agents are distinguished from the inferior agents to reduce adverse selection by such a mechanism [8, 9]. However the contract inevitably induces agents to mimic portfolios with each other, leading to herd behavior in the innovation adoption. Under the constraint conditions, herd behavior is a rational, utility-maximizing choice for agents.

Therefore, the key to solve herd behavior of innovation adoption is to design an effective incentive and restraint mechanisms by which agents share benefit from successful decision and responsibility of mistakes by avoid herd behavior and adopting innovations. As a result, agents' interest and owners' interest go in concert in the long tem.

3 Evolutionary Game Analysis on Herd Behavior in the Decision-Making Process of Innovations Adoption

3.1 Evolutionary Game Theory

Evolutionary game theory is based on the limited rationality of individuals and regards groups as the key research object. It pointed that individuals' behavior in reality is not optimal and individual decision-making is to achieve by imitating and learning others during a dynamic process [10]. The theory basically looks at group behavior adjustment process from an evolutionary view of "survival of the fittest". The general process of evolution includes two possible evolutionary mechanisms: selection mechanism and the mutation mechanism. Investors to develop an innovation decision-making have two characteristics as follows:

(1) Bounded rationality. Participants are presumed rational in many game models. However, investors are far from full rational in reality. This is partly due to speculative factors on one hand, and information asymmetry of innovations adoption on the other hand [5].

(2) Ability of repeated learning and to adjusting strategies. The process that investor make a decision of innovation adoption and obtain the expected profit cannot be completed through one transaction. Implementation of innovation adoption is a long-term repeated process by which investors learn and adjust their investment strategies back and forth to adapt to market changes.

3.2 Brief Introduction to the Model

Frogs-croaking game model in evolutionary games is an appropriate method of analyzing herd behavior among investors who make decisions of innovation adoption. If all investors do not analyze by themselves, any decision of adopting technological innovation will be extinguished in the entire market. As a result, the gains equals to zero generally. If one of a business makes a decision of adopting innovations based on the independent analysis, then the benefit increases to M apart from the cost of C and the total income for the adventurer equals to M-C. If another enterprise follows the example, he will gain an income of N from herd behavior. If two companies make the same decision of adopting innovations based on independent analysis respectively, both will obtain a gain of R while they have to pay the cost of C, the total income for both as R–C. The game matrix is shown in Fig. 1.

(1) R>M: it means that if there is an imitator, the benefit of innovation adopter will decrease.
(2) M>N: it means that because the follower enters the market later, his benefit is smaller than the forerunner.
(3) C>0: it is the cost of making a decision independently.

3.3 Game Analysis

Suppose that in the early time, participants who making decisions independently occupy a ration of x in the total, participants who imitate other occupy a ration of $1-x$ in the total. The Replicator Dynamic Equation is shown as follows:

$$\frac{dx}{dt} = x(1 - x)[x(R - C - N) + (1 - x)(M - C - 0)] \tag{1}$$

	Participant 2	
	Independent decision	Imitation decision
Participant 1 — Independent decision	(R-C, R-C)	(M-C, N)
Participant 1 — Imitation decision	(N, R-C)	(0, 0)

Fig. 1. Game matrix of herd behavior in innovation adoption

We find 3 possible stable points for Eq. (1), they are $x^* = 0, x^* = 1, x^* = \frac{M-C}{N+M-R}$

(1) If $0 < \frac{M-C}{N+M-R} < 1$. 3 points mentioned will be in the efficient range of $0 \leq x \leq 1$.
That is $M - C > 0$ and $N > R - C$. Suppose that $F(x) = \frac{dx}{dt}$, then
$F'(x) > 0, F'(1) > 0$, and $F'\left(\frac{M-C}{N+M-R}\right)$. It is obvious that $x^* = \frac{M-C}{N+M-R}$ is the
Evolutionary Stable Strategy. Under this condition, firms making decisions
independently occupy a ratio of $x^* = \frac{M-C}{N+M-R}$ in the total while firms imitating
others occupy a ratio of. $1 - x^*$ in the total. Therefore, to some extent, herd
behavior is shown in the entire market (Fig. 2).

(2) If. $\frac{M-C}{N+M-R} < 0$ We find 2 possible stable points for Eq. (1). They are
$x^* = 0, x^* = 1$. It is obvious that $x^* = 0$ is the Evolutionary Stable Strategy
because $F'(x) < 0$ and $F'(1) > 0$. Under this condition, no firm makes decisions
independently or adopt any innovation, Nash Equilibrium equals to (0, 0). The
condition of $M < C$ means that the cost of adopting an innovation is too high to
shoulder for any firm. Even if in the early time, there are a few firms who make
decision independently, they will give up because of the cost higher than the
benefit. Therefore, herd behavior becomes the only choice in the market (Fig. 3).

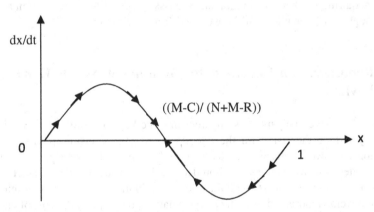

Fig. 2. Frogs-croaking evolutionary game $(0 < (M-C)/(N + M-R) < 1)$

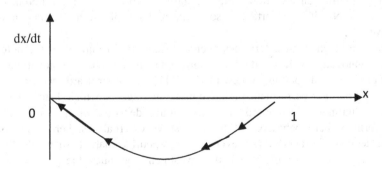

Fig. 3. Frogs-croaking evolutionary game $((M-C)/(N + M-R) < 0)$

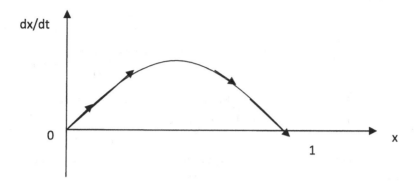

Fig. 4. Frogs-croaking evolutionary game ((M−C)/(N + M−R) > 1)

(3) If $\frac{M-C}{N+M-R} > 1$. We find 2 possible stable points for Eq. (1). They are $x^* = 0, x^* = 1$. It is obvious that $x^* = 1$ is the Evolutionary Stable Strategy because $F'(0) < 0$ and $F'(1) > 0$. There is a perfect Nash Equilibrium under this condition, (R–C, R–C). All firms make decisions independently to adopt innovations. What we could learn from this phenomenon is that if both the cost of acquiring market information and the risk of adopting an innovation are small enough, herd behavior will be expelled from the market (Fig. 4).

4 A Remuneration Contracts to Circumvent Agents Herd Behaviors

Based on the above analysis, it is apparent that the key to avoid herd behavior is to effectively reduce the risk with the agent who has the decision-making power of innovation adoption. Entrepreneur as a rational individual has to pay a cost when making a decision as well as to shoulder the loss brought by the uncertainty of innovations. To make it worse, with the knowledge in making a decision of innovation adoption develops more and more tacit, systematic and complex, the cost of enterprise innovation rockets at the same time. Even if the innovation is successful, benefit created by innovation adoption varies because entrepreneurs have different control power. Therefore, it is important to solve the problem of entrepreneurs' innovation incentive.

Incentive mechanism determines agents' behaviors. Incentives to promote technological innovation will inevitably encourage agents to prefer to innovations, R&D of new products and opening up new markets [11]. Agent's reward mechanism in the past depend on their performance relative with other agents, which distorted effective incentives and induced agents to choose the similar decisions of innovation adoption in the form of herd behavior. With the positive externalities, herd behavior is a rational decision for agents. However, in a competitive market, such selection will generate negative externality that if the investment continues, the profit of the firm will decrease. In order to avoid the losses from herd behavior in innovations adoption, principals hope that agents can acquire most information to make an optimal decision independently.

Therefore it is urgent that a remuneration system should be established to eliminate herd behavior in innovation adoption. The new contract model should be constituted of three parts: a fixed risk-free return, a performance-related return, and a return according to agent's behavioral risk (adopting an innovation or follow prior examples). Such a compensation contract allows for an opportunity cost higher than the expected profit for emulating other's decision, thus serving to circumvent agent's herd behavior when making a decision of innovation adoption.

5 Case Study: China's E-commerce Business Model Innovation Adoption

5.1 Development of E-Commerce in China

(1) A seed stage of 1999 to 2002
     According to statistics released in 2000, there are only 10 million Internet users in China at that moment. Applications for users at the stage included e-mail and news browser. Under the infancy of e-commerce environment, the few e-commerce firms survived apart from a number of early Internet users are developed.
(2) A rocketing stage of 2003 to 2006
     Several e-commerce giants in China has become a hot topic in the Internet arena, including Dangdang.com, Joyo, Alibaba, HC, Taobao, Which overwhelmed the entire communications and network world. The development of network applications has far-reaching impact on traditional business.
(3) An overall development from 2007
     E-commerce is more than at world of Internet companies. Numerous traditional businesses and capital flow into the area of e-commerce development in China will reach new heights in the near future. A new era of continuous integration of a physical world and virtual world is before users of billions.

5.2 Taobao and C2C, B2C E-Commerce Business Models in China

In 2004, Ma Yun spent 100 million RMB building Taobao.com and scored a 50 % share of C2C market. By 2005, C2C market was split by Taobao and eBay while others, such as Yahoo, Sina, were lagged behind far away. Later, Dangdang.com, a powerful B2C company in China announced its cross-field business entry into the C2C market. With its overwhelming strength in B2B, Taobao defeated Dangdang ultimately in C2C market easily. By 2007, Taobao has occupied a share of 80 % of C2C market.

Although the prospect of B2C was glooming before 2006, Jingdong Mall entered this market as the second heavy-weight opponents of Taobao. With a strategy "small profit and quick return" for 3C products, which was welcome by VC bigwigs at that time. And then B2C market began to boom Where the VANCL, Mbaobao.com, Moonbasa and other star-ups entered this market. In order to embrace the challenges from opponents as well as the new requirement of c-commerce, Taobao introduced a brand new branch as B2C platform: Taobao Mall in April 2008. As expected, with the complete eco-system of e-commerce from Alibaba Group, Taobao Mall is growing much more rapidly than others.

5.3   Latest Innovations of Alibaba Group

At June 16, 2011, Taobao was divided into three: Taobao, Taobao Mall, and eTao in order to let the huge Group "Forever Young". eTao, the first price-comparison website in China, not only provides a comparison business, but also show the unique value of each business from service, payment, logistics. eTao is seeking close cooperation with the mature B2C market. By the end of 2011, eTao has developed more than 6,000 partners, including new egg.com, VANCL who occupied top 10 in B2C market.

5.4   Lessons from Taobao

(1) Competition Situation. According to the statistics, in 2008, with a 18 % market share, Jingdong Mall became the largest independent firm in China's B2C businesses and occupied 47.6 % of the 3C online retail market, followed by Joyo with 15.4 % and Dangdang with 14.8 % respectively. However, difficulties are temporal for Taobao.
In 2011, Taobao has defeated three competitors, leho.com, eBay and paipai.com, hit No. 1 in C2C market while Taobao Mall ranked top 1 in B2C with a market share of 48.5 %, followed by Jingdong Mall with 18.1 %.

(2) Entrepreneurship: Taobao's success can not be separated from Ma Yun. Ma Yun concluded Taobao's secret of success that are: dream, learning and insistence. Mr Ma believes Internet will direct the change in China. China is an ideal place to develop e-commerce and the key to e-commerce development is to let e-commerce operators rich. In addition, Ma Yun pointed out dream will become a pain without insistence. Thirdly, Ability to learn is essential to the success of Alibaba Group. Last but not the least; Ma said that firm leader should choose a right direction. Fortunately, Alibaba's choice is looking-forward and sensible because e-commerce is the direction not only in China but also the entire world.

6   Conclusion

Innovation has become a key native power of firm development. A business agent is responsible for decision-making of innovation adoption. Herd behavior in the innovation adopting induces enterprises to miss opportunities for innovation and development, and then market bubbles and repeated investment appear. Improvement in incentives mechanism is a good method to avoid herd behavior, encourage entrepreneurs to make their own decisions, and maximize the interest of shareholders. Therefore, a diversified compensation structure, such as the one in Paragraph 4, should be designed to give the residual claim to agents to enable agent's to maximize their own utility as well as firm's interest from innovation activities at the same time.

References

1. Wu, Z., Yang, W.: Research o formation mechanism of herd behavior based on asymmetric information. Mod. Manage. Sci. **25**, 50–53 (2006)
2. Guo, Y., Su, X.: Empirical study on firm's decision-making of innovation adoption. Manage. Rev. **17**, 46–53 (2005)
3. Shen, M., Song, J.: A study of remuneration-based managerial herd behavior. China Econ. Q. **7**, 1013–1028 (2008)
4. Feng, B.: The game theory analysis on herding behavior in security market. J. Harbin Univ. Commer. (Soc. Sci. Ed.) **9**, 48–51 (2007)
5. Wu, Y., Liang, J.: Herd behavior and bounded rationality. Acad. Exch. **15**, 116–118 (2008)
6. Bikhchandani, S., Hirshleifer, D., Welch, I.: Learning from the behavior of others: conformity, fads, and informational cascades. J. Econ. Perspect. **12**, 151–170 (1998)
7. Bikhchandani, S., Hirshleifer, D., Welch, I.: A theory of fads, fashion, custom, and cultural change as informational cascades. J. Polit. Econ. **100**, 992–1026 (1992)
8. Abhijit, V.B.: A simple model of herd behavior. Q. J. Econ. **107**, 797–817 (1992)
9. Pi, T., Zhao, T.: Herd behavior review. J. Yunnan Univ. Finan Econ. **21**, 19–23 (2005)
10. Xie, S.: Economic Game. Fudan University Press, Shanghai (2010)
11. Duan, W., Gu, B., Whinston, A.B.: Information cascades and software adoption on the internet: an empirical investigation. MIS Q. **33**, 23–48 (2009)

Applying Data Mining to the Customer Relationship Management in Retail Trade

Yao Han$^{(\boxtimes)}$, Feng-zhao Yang, and Gang Wu

Jiangsu Provincial Key Laboratory of E-Business,
Nanjing University of Finance and Economics, Nanjing 210003, China
Hanyao1993@yahoo.com.cn

Abstract. Customer Relationship Management (CRM) has attracted much attention due to the high competition in the retail trade. Meanwhile, data mining has great potential to improve the effectiveness of CRM, since the data volume of customers is drastically increasing. In this paper, we firstly propose a logic structure of the retail CRM system, and thus discuss several data mining techniques, i.e., classification, clustering, association analysis, for CRM. Last but not the least, several sub-fields of CRM that are suitable to the application of data mining techniques are discussed. As can be seen from this paper, the retailers must strengthen the application of data mining technology research in order to support the management decision-making and improve the level of information management.

Keywords: Retail · Customer relationship management · Data mining

1 Introduction

Retailing is one of the industries that are developing most rapidly in the last century. Nowadays, retailing becomes an important pillar industry of national economy. The development of retailing depends considerably on the technology progress, especially information technology (IT). Since 1960s of last century, after the revolutions of organization and management, the revolution of technology was raised. Many techniques include IT, Web and electronic commerce (EC), were applied widely in retailing which reduced the costs and raised efficiency for retailers. The development of retailing was supported greatly by technology progress.

Today, the technology of information management is adopted by most of big retailers which is typified of Wal-Mart who has the management information system based on a global communication network that connects all chain shops and suppliers. Wal-Mart gets hold of competitive superiority by the application of IT and Web and became the biggest retailer, also biggest enterprise in the world.

Based on IT and Web, the application of CRM was also developed in retailing. CRM is a system and soft for gather, storage and analysis of customers' information. CRM could support the management the customers' information as a kind of resource and the decision-making in the business [1]. CRM have been implemented in some retailers as Wal-Mart and gets distinct effects [2].

S. Zhou and Z. Wu (Eds.): ADMA 2012 Workshops, CCIS 387, pp. 116–124, 2013.
DOI: 10.1007/978-3-642-41629-3_10, © Springer-Verlag Berlin Heidelberg 2013

Comparing the developed countries, the informationization and electronic commerce are also at a phase of beginning in China [3]. While many retailers have the management information system (MIS), it is only used in generic management as stocking, selling and accounting etc. A great lot information can not be used availably and transformed resource and wealth. Thus it is very necessary for retailers to implement CRM in order to advance the informationization and raise the power of competition [4].

2 Retail Relationship Customer Relationship Management and Data Mining

The retailing is a business which sells goods and services to the ultimate customers. The retailers do not produce products generally, but purchase goods then sell them to customers. The retailing divided into some different retail formats as department store, supermarket, monopoly store, convenience store etc. Most of the modern retailers take the model of chain operation that headquarters manage purchase, inventory and delivery and the chain stores manage sell and market.

The kinds of commodities sold by retailers are more than thousands upon thousands and their structure is very complex. There are a large numbers of data of operation should be processed which are the base of unified management by headquarters. If these data could be translated into knowledge and resource by CRM, the level of operation and management would be enhanced.

The customers of retailers almost are individual which are numerous and scattered and complex so the retailers can not take one-to-one marketing to everybody. Generally, the customers of retailers buy small quantity goods every time but buy goods frequently. Some people would be loyal to a certain store and frequently buy goods there. Discovering the loyal customers and maintaining the relationship with them, that is an important content of retail CRM.

The stores of different retailers cluster in the area of shopping center in order to attract maximum customers by the clustering effect, however the clustering of stores would bring on sharp competition and high marketing cost. If implement CRM, the retailers could cultivate and keep their loyal customers instead of price and promotion competition.

Actually, the early grocers have already adopted the method as like CRM. The shopkeepers have remembered of the habit and require of every familiar customer and gave them individual services. Then the kinds of merchandises and the amount of customers were increased greatly afterward, so the department stores could not give each customer one-to-one services as like grocers before. Now, the progress and application of IT and Web give strong support to one-to-one marketing for retailers who could serve everybody individually as like the grocers.

Of course, the role of CRM system is not limited to record the customers' data, its more important function is for business management to provide a scientific basis, to help enterprises to find out the rules of customer service from large amounts of data and support decision-making of business management. Obviously, without data mining, this function of CRM will not be able to achieve.

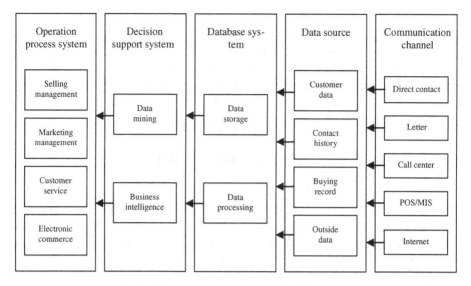

Fig. 1. The logic structure of retail CRM system

According to the operation logic and ingenerate relations of CRM, the system of retail CRM could be compartmentalized into five logistic layers as Fig. 1. As shown in Fig. 1, the decision support system is the CRM system the most important application part [5].

The main function of decision support system is analysis of the data in warehouse, using data mining and business intelligence tools from large amounts of data to extract useful knowledge of the customer, and guide to the retail business activities. Therefore, based on the data mining and business intelligence is a core technology of retail customer relationship management. Only to the customer data for data mining, the customer information could be translated into customer knowledge that is useful to retailers who want to better achieve the effective customer relationship management. Therefore, data mining for retail customer relationship management has important meaning and action.

3   The Data Mining Methods in Retailer Customer Relationship Management

Data mining is a kind of decision-making process, the basic idea is that to find and extract hidden valuable information from the large amounts of data in the database, help decision makers look for potential relevancy, find ignored factors, so as to grasp the laws, predict trends in the data, in order to scientific decision [6]. The data mining method has many kinds. The main methods which commonly used in the retail customer relationship management are prediction, classification, clustering and association analysis.

3.1 Prediction

Forecast is generally made according to the forecast analysis model. Predictive models usually assume that some phenomenon (the dependent variable) occurs for other phenomena (the independent variables) arising, or changes with other phenomena. A stable relationship exists between independent variable and dependent variable. In this way, the future status can be protected based on the known data. In data mining, constructing a prediction analysis model usually aims to detect the customer reflection to a specific marketing activity.

Logistic regression is a basic method that is most commonly used in the prediction that are used to construct the quantitative relation between the target variable (the dependent variable) and one or more predictor variables (independent variables). Usually, a logistic regression can be used to predict two or more than two kinds of results, as like the customer respond to the promotion or not. The typical logic regression model can be expressed in a formula for:

$$L_n \frac{p}{1-p} = \beta_0 + \beta_1 X_1 + \beta_2 X_2 + \ldots + \beta_n X_n \tag{1}$$

where p indicates the probability of target occurrence and $1 - p$ indicates the probability which target event does not occur, $\frac{p}{1-p}$ represents the probability of target event. A logistic regression model is a linear function regarding the natural logarithm of the odds as a series of independent variable. According to this principle, data mining software can make a model based on historical data and predicts the customer behavior (such as the response for a advertising).

3.2 Classification

Classification is the segmentation of target object data based on the argument sequence, all of the target objects are divided into different groups, Considerable heterogeneity exists among groups but great homogeneity shows in each group. Through the classification, the relationship can be found out between the influencing factors and the target event and make the anticipation about customer behavior.

Decision tree is a common method in classification which tests data sample from the root node, then the data samples are divided into different sample subsets according to the different results, each subset of data sample is a subset of nodes that corresponds to a classification. Constructing a decision tree was designed to identify the relations between attributes and categories, in order to predict the category of future unknown record.

For example, some retailers issued 10000 mail advertising, the response rate is 4.6 %. Then two variables were detected with the Gender and age, the results as shown in Fig. 2.

Decision tree analysis shows, male customer above 40 years old response rate levels is much higher than those under the age of 40; and the response rate of female customer below 40 is significantly higher than those over 40; male customer above 40 years old response rate is higher female customers blow 40. According to this

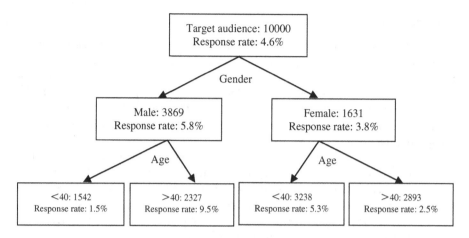

Fig. 2. Application of decision tree

result, the retailer would face mainly to more than 40 years old male customers, then the age of 40 female customers in similar promotional activities. It can improve the response rate and reducing the cost of marketing.

3.3   Clustering

Clustering aims to divide the database records into different groups or subsets on the certain attributes, records inside each group are similar in identifying attribute, but big difference exists in records among different groups [7]. This function of clustering can be used for customer segmentation.

When using clustering model in the retailing, customers will be divided automatically into different types, and then according to the different types of customer purchase behavior and demand characteristics, different marketing strategies are adopted [8].

For example, 10000 customer records of a store in the database, according to gender, age, marital status, education, occupation, income, average consumption, purchase duration and so on, are clustered into four clusters by clustering methods according to the natural relationship of data with some certain distance standards, then according to customer value the clusters are marked the class tag as A, B, C, D. The number, volume ratio and the contribution to the enterprise of each cluster are shown in Table 1.

From the analysis result, the customers of class A and class B account for only 21.6 % of total number of the customer, but the contribution ratio is 71.4 %; while customers class C and class D accounted for 78.4 %, but the contribution ratio only is 28.6 %. Accordingly, the customer of class A would be identified as optimal customer, class B as the main customers, class C as the general customers, while class D as small customers, and the store would makes different marketing strategies and offer different services for different customers.

Table 1. Clustering results

	A	B	C	D
Number	842	1318	2769	5071
Volume ratio (%)	8.42	13.18	27.69	50.71
Contribution ratio (%)	46.32	25.08	20.14	8.45

3.4 Association Analysis

Association analysis is to find the relationship of different items in a same event. It can be formally described as: $I = \{I_1, I_2, \ldots, I_m\}$, m is the collection of different item, D is the collection for I on event, every event in D includes a number of project I', and I' belongs to I, then the association rules expressed as $X => Y$, X and Y belongs to I, and $X \cap Y = \emptyset$. X called rule premise, Y is the result. For each rule, the support and confidence should be calculated. The rule which support and confidence are both more than the given thresholds of minimum support and minimum confidence is considered to have a reference value, and is included in the analysis results.

Association analysis which is one of the main functions of data mining can be widely used in marketing, One of the applications is to study on customer purchasing behavior and explore the behavior patterns of customers in the commodity purchase. In a store as an example, there are recorded data of customer purchase household appliances category within the scope of certain time as shown in Table 2.

According to the data in the table above, a correlational analysis table is constructed as shown in Table 3, the data in the table show the dealing times that customers purchase more than two kinds of goods or purchase a separate commodity.

Based on the correlation analysis to the data in the table the support and confidence are calculated. The meaning of Support is the percentage of dealing times of the shopping basket contains both commodities of association rules, and that is also the percentage of supporting this rule, equivalent to the joint probability. Such as "refrigerator-washing machine" support is $Sup(X1-Y1) = P(X1|Y1) = 32/100 = 32\%$. Confidence is all to buy one commodity dealing, while purchasing

Table 2. Data of the purchase types of home appliance

Dealing item	Dealing number	Dealing item	Dealing number
Refrigerator, washing machine	20	Microwave oven, dryer	22
Microwave oven, washing machine	8	Refrigerator	8
Refrigerator, microwave oven, washing machine	12	Microwave oven	10
Washing machine	6	Electric oven	3
Dryer	4		
Refrigerator, dryer	7	Total	100

Table 3. Correlation analysis

	Washing machine X_1	Dryer X_2	Separate purchase	Total
Refrigerator Y_1	32	7	8	47
Microwave oven Y_2	20	22	10	52
Electric oven Y_3			3	3
Separate purchase	6	4		
Total	58	33		

Table 4. Support and confidence

	Support		Confidence	
	Washing machine	Dryer	Washing machine	Dryer
Refrigerator	0.32	0.07	0.68	0.15
Microwave oven	0.20	0.22	0.38	0.42

another commodity, this is a conditional probability. Such as "refrigerator-washing machine" confidence is $Con(X1\text{-}Y1) = P(X1|Y1)/P(X1) = 32/47 = 68\%$. All the calculation results are shown in Table 4.

According to the minimum threshold of support and confidence which were set, some items that are more than the minimum threshold will found out. These items are considered as commodity combination of correlation rules. Assumed a minimum support threshold and the minimum confidence threshold were 0.3 and 0.5 respectively in this case, only the support and confidence of "refrigerator-washing machine" is more than the threshold. It can be concluded that there is an association rules between refrigerator and washing machine perchance that mean the customer may buy the refrigerator and washing machine both.

4 Application of Data Mining in Retail Customer Relationship Management

There is a wide range of applications of data mining in the retailing, it can be used on the customer, goods and store management, can also be used in marketing management and knowledge management. Through the application of the technology of data mining, which is stored in a database of customer information to the resource of the enterprise, for the retailers to support effective customer relationship management and improve operational efficiency.

4.1 Customer Management

The methods of data mining could be used in customer management firstly. According to the famous Paredo Law, 80 % profit come from 20 % customers. These customers, who are few but valuable, should be main object to serve in order to keep and solidify the relationship with them and prolong their customer lifecycle. Data mining could help retailers to separate valuable customers from others and support the decision-making of service mix.

4.2   Merchandise Management

The Paredo Law is also the same with commodities selling, 80 % profit come from 20 % sold merchandises. The retailers should know what commodities are profitable and what are not so that the retailers could adjust and optimize constantly the structure of commodities. Data mining could help the retailers discover the most profitable commodities and lowest value commodities from whole, then adjust commodity structure and arrange commodities storage and purchase.

4.3   Chain Store Management

The business performance of retailer's chain stores has difference, this difference may be due to branch of the internal reasons as operation and marketing, may also be due to external environmental factors as location and customer groups. The relations between the operational performance and the differences of operation, marketing or location, customers could be fined by data mining so that headquarter could give the chain stores correct evaluations and feasible guidance.

4.4   Marketing Management

Data mining is more useful in marketing management. By data mining, the retailers could analyses the characters of different customers' demand, fine the model or rule of customer behavior and take different marketing strategies to each of status. Data mining also could be used in the result analysis of marketing strategies that could help the retailers to evaluate the marketing effect exactly and improve on marketing strategies onwards.

4.5   Knowledge Management

An important function of data mining is the extraction of information that is useful but implicit, unknown and potentially from a large, incomplete, noisy, blurred, practical random data which after refining can be transformed into enterprise knowledge. Therefore, data mining is essential tools of retailer knowledge management also. Through data mining, the retailers could be better to detect, format and spread customer knowledge and improve customer service level and market competitiveness.

5   Conclusion

With the rapid development of information technology, customer relationship management has become the main means of improve customer satisfaction and loyalty then to achieve a competitive advantage. The retail industry is facing the end user, their amount is huge and structure is complex. If retailers want their customers to be managed effectively, they must carry on processing to the massive customer data to extract useful information with the aid of various data analysis technology, especially the technology of data mining, thus to determine the customer's value, predict customer behavior and extract customer knowledge. The application of data mining in the retailing has very momentous significance, it makes the retailers to carry out more practical and knowledge-oriented decision making, help them to better customer management, merchandise management, store management, marketing management and knowledge management, and realize the management informationization at the higher level.

Acknowledgments. This Research is supported by A Project Funded by the Academic Program Development of Jiangsu Higher Education Institutions (PAPD).

References

1. Mendoza, L.E., Marius, A., Pérez, M., Grimán, A.C.: Critical success factors for a customer relationship management strategy. Inf. Softw. Technol. **49**(8), 913–945 (2007)
2. Anderson, J.L., Jolly, L.D., Fairhurst, A.E.: Customer relationship management in retailing: a content analysis of retail trade journals. J. Retail. Consum. Serv. **14**(6), 394–399 (2007)
3. Sha, Z.: The application of customer relationship management in retailing. Today Hubei (Theory Edition) **1**(5), 240–241 (2007)
4. Gao, Y., Xu, W., Yuan, L.: Analysis and structure of customer relationship management system based on retailing. Sci. Technol. Manage. **Sum 24**(2), 40–42 (2004)
5. Gu, M., Chen, J.: Research of the retail trade CRM based on DSS. J. Suzhou Inst. Silk Text. Technol. **25**(6), 25–28 (2005)
6. Sun, Y., Zhao, P.: The retail CRM based on data mining. Mark. Modernization **Sum 504**(15), 6–7 (2007)
7. Yuan, Q.: The application of data mining in retailing. China Comp. Commun. **Sum 132**(12), 52–55 (2009)
8. Cao, J., Wu, Z., Wu, J., Liu, W.: Towards information-theoretic K-means clustering for image indexing. Sig. Process. (2012) http://dx.doi.org/10.1016/j.sigpro.2012.07.030

Informatization in Grain Circulation of China: A Survey

Baifei Tang[(✉)]

School of Economics and Management, Beijing Jiaotong University, Beijing, China
baifeitang@chinagrain.gov.cn

Abstract. Although the grain industry is a traditional businesses, grain storage, distribution, trading and sales model have occurred remarkable change through the application of information technologies. In this paper, the representative technologies applied in the grain circulation are first introduced and analyzed. Then, the existing problems in grain informatization are discussed based on the statistics data. Finally, the develop trend and future researching fields are analyzed.

Keywords: Informatization · Grain circulation · Storage · Transportation

1 Introduction

Along with development of information technology, various forms of applications have been carried out in the grain industry. The study of different types of applications of information technology and information integration technology have been involved. With the IT applications in the grain circulation management gradually, a new situation of grain circulation management is coming.

The state department for grain administrations attach great importance to IT application and development in the grain industry [1, 2]. They carry out a series of top-level design about the standard of grain information coding. There is a medium- and long-term planning about the application of the computer technology in the grain industry. And combined with the actual need of the grain, the national grain reserve information management systems for the preparation and grain condition monitoring and control and the development of standardized software of the mechanical ventilation control systems have been presented, which laid a strong foundation for the grain information construction and the development of information technology.

A systematic demonstration of the application of information technology in the grain industry has been carried out during the "Eleventh Five-Year" period. To ensure grain security and to develop grain industry, the scientific and technological progress which marked by the rapid detection of food safety in grain storage security, the grain traceability and the grain quality controlling technology have played an important supporting role [3, 4]. The grain circulation management network system has been constructed in Shanghai, Zhejiang, and Ningxia, according to their local conditions; The wireless system of grain condition monitoring and controlling are used by the refined grains emergency library in Suzhou National Grain Reserve; The solar low

S. Zhou and Z. Wu (Eds.): ADMA 2012 Workshops, CCIS 387, pp. 125–129, 2013.
DOI: 10.1007/978-3-642-41629-3_11, © Springer-Verlag Berlin Heidelberg 2013

temperature rice storage technology has been applied exemplarily in Yangzhou Depot; China Grain Reserves Corporation starts the application of the internal grain storage business management system; The informatization also be included in the design phase of the project by the Dongguan library [5].

Grain administrative units need to know the actual situation on the grain circulation timely and accurately. Therefore grain storage information, price information, flow rate information, traffic information should be reflected promptly and accurately [6, 7]. The administrative units should also to develop appropriate managing policies. Grain condition monitoring and control system as the application of information technology in the grain industry change the form of the grain barn management. So the development of information technology will be able to upgrade the level of the grain administration. The mode is changed from the original plan dispatching, information statistics lag, to the on-site visual supervision and management. Now he flow of information can be traced.

Information technology is a high-tech development in China. The food industry application of information technology means the transformation and upgrading of traditional industries with the use of advanced technology achievements. The scattering and natural grain acquisition, distribution, storage, marketing and other aspects of the lack of organization will be connected and got together by the application and promotion of information technology. It not only enhances the efficiency of the grain circulation, but also achieves the transparent information and reduces the risk of grain circulation system. It also brings a revolutionary change on the grain enterprises, managing mode and improves the level of management significantly. Information technology enables the information about the operating of grain library becomes real, fast and accurate [8]. The operating process will be automation and controlled precisely. It helps to improve the operational efficiency and reduce the jobbery phenomenon.

2 Existing Problems

2.1 Separated Testing Platform

The development of information technology in the grain industry is related to the development of information technology closely. The development and innovation of information technology lead to the direction of development of information technology. But a few isolated information models cannot fully implement the role of information technology in promoting the development of grain circulation. It also can not have good effect on the promoting of the efficiency of circulation and the capacity of management. Meanwhile the initial investment costs is high on the information technology [9], but the direct benefits are slowing. The administrative department of grain and grain enterprises will have a high pressure on the economy. And the necessity of informatization in the grain industry will be obliterated. Furthermore, it will restrict the pace of the development of information technology in the grain industry.

2.2 Lack of Standards

The information construction of the grain industry is based on the specification of information technology. The main agency of the construction draw up the specifications, the coding of the exchange of information, the system architecture according to their own need because there is no uniform of guidance [10]. It will lead to the information construction system of all main agency cannot be combined organically and information silos. Grain circulation information will be collected only in the specific region and that information about exchange which is out of the region cannot be gathered. So it will lead to the information construction becoming difficult and information standards are not unified and various information systems are in their own way and increasing the phenomenon of repeating investment.

3 Methodologies

3.1 Unification of Different Resources

The exploration of the application of the Information technology in the grain warehouse management and the process of grain circulation proves that the information technology can not only to improve grain storage management efficiency, but also can to enhance the effectiveness of the grain circulation management and make the grain business management have the max benefits. To ensure the security of the grain and support the macro-control of grain rapidly and effectively, the grain industry should have a unified understanding of the application of information technology and seek the input of financial funds for grain information construction actively. In order to create a good atmosphere of the grain circulation modernization, the grain industry should also make the construction of grain circulation into the area of the basic construction.

3.2 Top-Level Designed Infrastructure

Grain circulation management is a systematical problem. It requires in-depth study of the relationship between information communication and grain circulation. The information related to management, operation, supervision, and circulation have to be integrated efficiently. Therefore, it is of great significance to protect the stability of grain circulation through the top-level design of the construction of information infrastructure. Scientific, rational and efficient construction program can not only fulfill the goals of information construction, but also guarantee the stability, safe, efficient of the grain circulation. It lays a solid foundation for the macro-control and the protection to emergency.

3.3 Grain Circulation Standardization

The application and development of IT technologies requires unified standards as a basis. The standardization of the grain storage information is very important for the development of information technology in the grain circulation. Therefore, the grain management authorities need to draw up uniform standards or General Clauses according to the existing results of the applying information technology with the means of the developing direction of information technology in the grain circulation. Its target is to achieve the comprehensive applications of the information technology in the grain circulation. The General Clauses about the informatization of the grain storage management should follow the condensing and applicable principle. It is necessary to study from successful modes which have been developed in some provinces, warehouse to form the General Principles which is applicable for the informatization of the industry-wide warehouse management, such as the model of the grain storage management information in Jiangsu. We should also research the General Rules of the informatization of the warehouse management gradually which can be used in the whole China's grain industry.

3.4   Demonstration Systems

The purpose of applying information technology is to improve the efficiency of the grain circulation, the security of the grain, and the efficiency of the grain supervision and management. It needs to increase the guiding role of the demonstration on the existing basis and continue to increase the investment to the information technology in the grain industry. It is required to develop the construction of information technology in the grain industry on the principle of unified, viable and efficient, to improve the efficiency of the managing work of the information technology and form a meaningful model in the grain industry, and to guide the national grain system based on the application of the information technology. Meanwhile, we should try our best to research the information technology to escalate and improve the accuracy and expand the detection field, reduce costs, increase durability, on the basis of difference requirements from individual provincial production. The finally target is to provide technical supporting to the informatization of grain industry.

4   Conclusion

The application and development of information technology brings opportunity for transforming the traditional industries to high technology in the grain industry. It is not only the way to improve the efficiency of the grain circulation and to ensure the grain security, but also the power to promote the change of the management mode. Therefore, in the context of information technology development, we have to seize the opportunity, conceive boldly, argue carefully, increase investment, and build the grain information system and the technological foundation to promote the large-scale development of the grain industry and to improve the efficiency of grain circulation.

References

1. Sun, X., Chao, O., Wu, G., Chen, S.: The innovative application of information technology in the grain storage and logistics enterprises. Cereal and Food Ind. **16**(5) (2009)
2. Yang, S.: Status and development into granary grain condition monitoring and control system, Storage Professional Branch of the Chinese Cereals and Oils Association. In: 2004 Annual Conference Proceedings (2004)
3. Abelson, H., Allen, D., Coore, D., Hanson, C., Homsy, G., Knight Jr., T.F., Nagpal, R., Rauch, E., Sussman, G.J., Weiss, R.: Amorphous computing. Commun. ACM **43**(5), 74–82 (2000)
4. Floerkemeier, C., Lampe, M.: Issues with RFID usage in ubiquitous computing applications. In: Ferscha, A., Mattern, F. (eds.) PERVASIVE 2004. LNCS, vol. 3001, pp. 188–193. Springer, Heidelberg (2004)
5. Tim, K.: Implementing physical hyperlinks using ubiquitous identifier resolution. In: Proceedings 11th International Conference on World Wide Web, pp. 191–199. ACM Press (2002)
6. Li, X.Q.: Access 2000 Applications In the Granary Information Technology Management. Technology Communication on Grain and Oil storage (2) (2006)
7. Zhang, H.: Using information technology to speed up the hair of our modern grain logistics exhibition. Grain Economy in China (10):40–44 (2003)
8. Cao, J., Wu, Z., Wu, J., Liu, W.: Towards information-theoretic K-means clustering for image indexing. Sign. Process. (2012)
9. Wu, Z., Cao, J., Wang, Y.: Dynamic advance reservation for grid system using resource pools. In: Altman, E., Shi, W. (eds.) NPC 2011. LNCS, vol. 6985, pp. 123–134. Springer, Heidelberg (2011)
10. Kimura, M., Saito, K.: Tractable models for information diffusion in social networks. In: Fürnkranz, J., Scheffer, T., Spiliopoulou, M. (eds.) PKDD 2006. LNCS (LNAI), vol. 4213, pp. 259–271. Springer, Heidelberg (2006)

Social Image Retrieval and Visualization

Effective Location-Based Image Retrieval Based on Geo-Tags and Visual Features

Yi Zhuang[1(✉)], Guochang Jiang[2], Jue Ding[3(✉)], Nan Jiang[4], and Gankun Zhu[2]

[1]College of Computer and Information Engineering, Zhejiang Gongshang University, Hangzhou, People's Republic of China
zhuang@zjgsu.edu.cn
[2]The Second Institute of Oceanography, SOA, Hangzhou, People's Republic of China
[3]Zhejiang Economic and Trade Polytechnic, Hangzhou, People's Republic of China
[4]Hangzhou First People's Hospital, Hangzhou, People's Republic of China

Abstract. With emergence and development of Web2.0 and location-based technologies, location-based image retrieval and indexing has been increasingly paid much attention. In the state-of-the-art retrieval methods, geo-tag and visual feature-based image retrieval has not been touched so far. In this paper, we present an efficient location-based image retrieval method by conducting the search over combined geotag- and visual-feature spaces. In this retrieval method, a cost-based query optimization scheme is proposed to optimize the query processing. Different from conventional image retrieval methods, our proposed retrieval algorithm combines the above two features to obtain an uniform measure. Comprehensive experiments are conducted to testify the effectiveness and efficiency of our proposed retrieval and indexing methods respectively.

Keywords: Social image · High-dimensional indexing · Probabilistic retrieval

1 Introduction

With emergence and development of Web 2.0 and location-based query, location-based image retrieval and indexing have been increasingly paid much attention. In the state-of-the-art location-based image retrieval methods [1, 2], geotag- and visual feature-based image retrieval has not been systematically studied so far.

Figure 1 shows an image and its related spatial information from google map. In this figure, different images have been annotated in the map, which is called *Geo-tagging*. Geotagging is the process of adding geographical identification

This paper is partially supported by the Program of National Natural Science Foundation of China under Grant No. 61003074, No. 61103229; The Program of Natural Science Foundation of Zhejiang Province under Grant No. Z1100822, No. Y1110644, Y1110969, No.Y1090165; The Science & Technology Planning Project of Wenzhou under Grant No. G20100202.

S. Zhou and Z. Wu (Eds.): ADMA 2012 Workshops, CCIS 387, pp. 133–142, 2013.
DOI: 10.1007/978-3-642-41629-3_12, © Springer-Verlag Berlin Heidelberg 2013

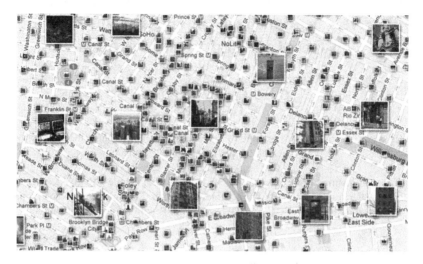

Fig. 1. Geo-tagged images

metadata to various media such as a geotagged photograph or video, websites, SMS messages, QR Codes [1] or RSS feeds and is a form of geospatial metadata. This data usually consists of latitude and longitude coordinates, though they can also include altitude, bearing, distance, accuracy data, and place names.

Geotagging can help users browse and watch the images with their location-specific information in an intuitive way. For instance, one can find images taken near a given location by entering latitude and longitude coordinates (see Table 1) into a suitable image search engine. Additionally, geotagging-enabled information services can also potentially be used to find location-based news, websites, or other resources [1]. Geotagging can tell users the location of the content of a given picture or other media or the point of view, and conversely on some media platforms show media relevant to a given location. Therefore, it is important to model an image with geo-tagging and visual features.

In most cases, people would like to get some result images with a specific location region they prefer to, which can not be achieved by the traditional semantic- or content-based search. Moreover, the retrieval effectiveness can be further enhanced if the visual features can be involved in this search process. Therefore, in this paper, we propose a composite-feature-based image retrieval method that combines *geo-tag* and *visual* features together. To optimize this retrieval process, we also propose two

Table 1. The images with geotagging information

Images	GPS
I_1	(30,50)
I_2	(10,150)
I_3	(60,20)

histogram-based query cost estimation schemes called a geotag histogram (*GH*) and a visual histogram (*VH*).

The primary contributions of this paper are as follows:

1. We present an effective location-based retrieval method by choosing combined features (i.e., *geotag* and *visual features*) of image.
2. We introduce a *cost-based query order selection* (CQS) scheme to optimize the query processing.
3. We perform extensive experiments to evaluate the effectiveness and efficiency of our proposed retrieval method.

The remainder of this paper is organized as follows. In Sect. 2, we provide the related work. Then, we give preliminaries of this work in Sect. 3. In Sect. 4, we first introduce the query cost estimation scheme. After that, a location-based image retrieval method is proposed. In Sect. 5, we report the results of extensive experiments which are designed to evaluate the efficiency and effectiveness of the proposed approach. Finally, we conclude in the final section.

2 Related Work

The research of image retrieval has been extensively studied during the two decades [3]. The QBIC [3] is known as a first system to support content-based image retrieval. After that, many prototype systems are built, such as VIRAGE [4], PHOTOBOOK [5] and MARS [6], etc. In all of these state-of-the-art CBIR systems, however, low level visual features such as color histogram, texture and shapes are only adopted without considering tag information.

As a new media type, the management of social image has been becoming a hot topic. According to the characteristics, Cui et al. [7] proposed a multi-feature-fusion-based social media retrieval and recommendation method, in which a improved inverted list is used to support the retrieval process. Siersdorfer et al. [9] provided a method for analyzing and predicting the subjective information of the social images, in which the SentiWordNet is adopted to extract the subjective features from the accompanying textual information. After that, a social recommender system is proposed [10]. Xin et al. [11] presented a prediction method for opinions and trends by discovering the potential correlation between social images through statistical analysis of user voting., opinions and trends, forecasts and machine learning techniques. In [8], Zhuang et al. proposed a social image retrieval scheme based on a *hypergraph spectral hash* model. This scheme, however, has not combined the tag and visual features together to support the probabilistic retrieval. Bu et al. [12] proposed a hypergraph-based social music recommendation method. Personalized information, however, has not been considered in this method. In this paper, we further study an effective image retrieval method by combing the geo-tag information and the visual features with probabilistic guarantee.

3   Preliminaries

First we briefly introduce the notations that will be used in the rest of paper (Table 2).

Table 2. Meaning of symbols used

Symbols	Notations
ψ	A set of images
I_i	The ith image and $I_i \in \psi$
n	The number of images in ψ
m	The number of reference images
$Sim\ (I_i, I_j)$	The similarity distance between two images
I_q	A query image user submits
$\Theta\ (I_q, r)$	A query sphere with centre I_q and radius r

Definition 1. *Geo-tag (GTag) can be modeled by a three-tuple:*

$$GTag :: \ <tID, mID, GPS> \qquad (1)$$

where

- *tID is the tag ID;*
- *mID is the tagged image ID;*
- *GPS is the location information of the mID-th image;*

4   Location-Based Composite Retrieval

In this section, to optimize the composite retrieval, we first introduce a cost-based query order selection (CQS) scheme in Sect. 4.1. Then we propose a location-based composite geo-tagged image retrieval (LCIR) method in Sect. 4.2.

4.1   Cost-Based Query Order Selection

For this location-based composite image retrieval, there are two query order schemes:

(1) *A geotag-based query is first performed, then a visual similarity query of the candidate images are conducted to obtain the result images;*

(2) *A visual similarity query is first conducted to obtain the candidate images, then a geotag-based query is performed to obtain the result images.*

For a large-scale image database, the above mentioned two query schemes have two different query costs. So we propose two cost-based query order selection schemes in which two histogram models are introduced.

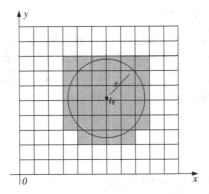

Fig. 2. An geo-tag histogram

• GeoTag Histogram Model

First, as shown in Fig. 2, we propose a geo-tag histogram (*GH*) model in which the space is partitioned into *m**m blocks. For each block, the probability that the geo-tagged images fall in this block is recorded.

Definition 2 (Geo-Tag Histogram, *GH*). *Geo-Tag Histogram (denoted as GH) can be represented by a two tuple:*

$$GH : <BlockID, Per>   \tag{2}$$

where

– *BlockID is the block ID;*
– *Per is the probability that the geo-tagged images fall in the block, formally denoted as: Per* $= \sum_{Ii \in BlockID} 1/|\Omega|$.

For example, for a query centered by I_q and r as a query radius, its affected blocks are shown by the shadow region. With the number of blocks increases, the appearance probability (*Per*) is near to that of the sum of the probabilities of the affected blocks.

Algorithm 1. Gtag query cost estimation

Input: Ω: the image set;
  T_q *(GPS.x, GPS.y)*: GPS information of I_q, ; R: spatial radius;
Output: P_1: the estimated query cost;
1. $P_1 \leftarrow 0$;
2. User submits a T_q and R;
3. the affected blocks are identified;
4. **for** each affected blocks B_i **do**
5. $P_1 \leftarrow P_1 \cup B_i.Per$;
6. **end for**
7. **return** P_1

- **Visual Feature Histogram Model**

Similarly, for the similarity query, we also propose a visual feature histogram model in which m images are randomly selected as reference images I_R.

Definition 3 (Visual Feature Histogram). *A visual feature histogram (VH) of reference image I_R can be represented by a triplet:*

$$VH : <rId, r, Per> \tag{3}$$

where

- rId *is the ID number of reference image;*
- r *is a sampling query radius;*
- *Per is the ratio of the number of candidate images that are obtained by a query sphere centered as the rId-th image and r as a radius of that of all images;*

Algorithm 2. Visual feature query cost estimation

Input: Ω: the image set; I_q: query image; r: query radius;
Output: P_2: the estimated query cost;
1. $P_2 \leftarrow 0$;
2. User submits a I_q, r;
3. find the nearest neighbor reference image(NNR) of I_q;
4. for a new hypersphere(HS) centered as $NNR(I_q)$ and r as radius
5. $P_2 \leftarrow HS.Per$;
6. **return** P_2

Based on the above two histograms, we can estimate and make a comparison of the query costs.

4.2   The Query Algorithm

In this subsection, we present the location-based composite image retrieval. The whole query processing can be divided into three stages: (1) *query submission* including query image I_q, query radius r, query geotag T_q and spatial radius R; (2) *query order selection*: according to the query criterion, firstly, obtain the proportion P_1 of the candidate images corresponding to T_q and radius (R) from *GH*. Secondly, find the nearest neighbor reference image (I_R) of I_q, then obtain the proportion P_2 of the candidate images corresponding to the query image (I_R) and query radius (r) from *VH*. If $P_1 < P_2$, we can choose the first query scheme (lines 5−10); vice versa (lines 12−18); (3) return the query result (line 19).

Algorithm 3. Location-based composite query

Input: Ω: the image set; I_q: query image; r: query radius;
$\quad\quad T_q$ *(GPS.x, GPS.y)*: GPS information of I_q, ; R: spatial radius;
Output: S: the query result;
1. $S_1 = S \leftarrow \Phi$;
2. User submits a I_q, r and *GPS* information;
3. the query order selection is conducted;
4. **if** $P_1 < P_2$ **then**
5.   **for** each image $I_i \in \Omega$ **do**
6.   **if** *dist(I_i.GPS,I_q.GPS)*<R **then** $S_1 \leftarrow S_1 \cup I_i$;
7.   **end for**
8.   **for** each image $I_i \in S_1$ **do**
9.   **if** *vSim(I_i, I_q)*<r **then** $S \leftarrow S \cup I_i$;
10.   **end for**
11. **else**
12.   **for** each image $I_i \in \Omega$ **do**
13.   **if** *vSim(I_i, I_q)*<r **then** $S_1 \leftarrow S_1 \cup I_i$;
14.   **end for**
15.   **for** each image $I_i \in S_1$ **do**
16.   **if** *dist(I_i.GPS,I_q.GPS)*<R **then** $S \leftarrow S \cup I_i$;
17.   **end for**
18. **end if**
19. **return** S

5 Experiments

In this section, we present an extensive performance study to evaluate the effectiveness and efficiency of our proposed retrieval and indexing method. The image data we used are from *Flickr.com* [13] which contains a set of the 50,000 geo-tagged images. We have implemented an online location-based geo-tagged image retrieval system (see Fig. 3) based on the Android platform [14] to testify the effectiveness of our proposed retrieval method. The retrieval approach is implemented in *C* language. All the experiments are run on a Pentium IV CPU at 2.0 GHz with 2G MB memory. As few research work has been touched on the composite retrieval by combing the geo-tag and visual features, so the baseline in the following experiments is a sequential scan without the CQS support.

5.1 Comparison of Two Histograms

In the experiment, we study and compare the estimation accuracy of the two histograms. We randomly generate 10 queries and compare two histograms. In Fig. 4, it is evident that the estimation accuracy of the gtag-based histogram is superior to that of the visual feature-based one. This is because the later one will result in a larger estimation region.

Fig. 3. A query interface of the retrieval system

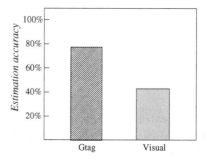

Fig. 4. Effect of reference images

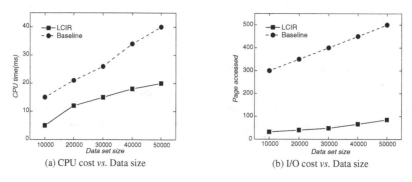

(a) CPU cost *vs.* Data size   (b) I/O cost *vs.* Data size

Fig. 5. Effect of data size

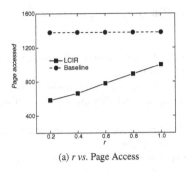
(a) *r vs.* Page Access

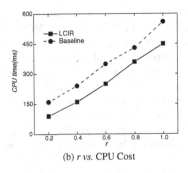
(b) *r vs.* CPU Cost

Fig. 6. Effect of *r*

5.2   Effect of Data Size

The experiment measures the performance behavior with varying number of images. Figure 5 show the performance of query processing in terms of CPU and I/O costs. It is evident that the CPU and I/O costs of the LCIR increases slowly as the data size grows. This is because the CQS scheme can effectively prune the search region and the computation cost of the candidate images is reduced accordingly.

5.3   Effect of *r*

In the final experiment, we proceed to evaluate the effect of *r* on the performance of a query process. Figure 6a and b both indicate that when *r* ranges from 0.2 to 1, the LCIR is superior to the retrieval method without the CQS in terms of page access and the CPU cost. The results conform to our expectation that the search region of the LCIR is significantly reduced and the comparison cost between any two candidate images is reducing as well.

6   Conclusions

In this paper, we presented a location-based composite image retrieval based on *geotag-* and *visual-features*. The prototype retrieval system is implemented to demonstrate the applicability and effectiveness of our new approach to image retrieval.

References

1. Zhang, J., Hallquist, A., Liang, E., Zakhor, A.: Location-based image retrieval for Urban environment. In: Proceedings of ICIP (2011)
2. Kawakubo, H., Yanai, K.: GeoVisualRank: a ranking method of geotagged images considering visual similarity and geo-location proximity. In: Proceedings of the 20th International Conference on World Wide Web, pp. 69–70

3. Flicker, M., Sawhney, H., Niblack, W., Ashley, J.: Query by image and video content: the QBIC system. IEEE Comput. **28**(9), 23–32 (1995)
4. Virage Inc. http://www.virage.com (2005)
5. Pentland, A., Picard, R.W., Sclarof, S.: Photobook: content-based manipulation of image databases. Int. J. Comput. Vision **18**(3), 233–254 (1996)
6. Mehrotra, S., Rui, Y., Chakrabarti, K., Ortega, M., Huang, T.S.: Multimedia analysis and retrieval system. In: Proceedings of the 3rd International Workshop on Multimedia Information Systems, Como (1997)
7. Cui, B., Tung, A.K.H, Zhang, C., Zhao, Z.: Multiple feature fusion for social media applications. In: SIGMOD Conference, pp. 435–446 (2010)
8. Zhuang, Y., Liu, Y., Wu, F., Zhang, Y., Shao, J.: Hypergraph spectral hashing for similarity search of social image. In: ACM Multimedia, pp. 1457–1460 (2011)
9. Siersdorfer, S., Minack, E., Deng, F., Hare, J.S.: Analyzing and predicting sentiment of images on the social web. In: ACM Multimedia, pp. 715–718 (2010)
10. Siersdorfer, S., Sizov, S.: Social recommender systems for web 2.0 folksonomies. In: Hypertext, pp. 261–270 (2009)
11. Jin, X., Gallagher, A.C., Cao, L., Luo, J., Han, J.: The wisdom of social multimedia: using flickr for prediction and forecast. In: ACM Multimedia, pp. 1235–1244 (2010)
12. Bu, J., Tan, S., Chen, C., et al.: Music recommendation by unified hypergraph: combining social media information and music content. In: ACM Multimedia, pp. 391–400 (2010)
13. http://www.Flickr.com (2010)
14. http://www.developer.android.com (2010)

HTML5 Based 3D Visualization of High Density LiDAR Data and Color Information for Agriculture Applications

Bo Mao[✉] and Jie Cao

Jiangsu Provincial Key Laboratory of E-Business,
Nanjing University of Finance and Economics, Nanjing, China
maoboo@gmail.com

Abstract. In this paper, an online visualization framework for agriculture LiDAR point cloud is proposed based on HTML5 technologies. In the framework, 3D data is transmitted with Websocket, and visualized with X3DOM that allows including X3D elements as part of any HTML5 DOM tree. There is no specific plugins required for the framework that extends the availability of the LiDAR data for agriculture related applications. The experimental results demonstrate the online visualization results of two types of LiDAR point clouds for the agriculture fields.

Keywords: Agriculture fields · Visualization · X3DOM · HTML5 · LiDAR

1 Introduction

Along with the development of remote sense technologies, we can monitor large agriculture areas in details. The remote sense data can be used to evaluate the growth status of crops and estimate the yields of fields. Many methods are proposed to identify nutrient deficiencies, diseases, water deficiency or surplus, weed infestations, insect damage, hail damage, wind damage, herbicide damage, and plant populations. Not only images but also 3D information about the agriculture can be generated in high density. LiDAR or Light Detection And Ranging technology can capture the shape information of objects with ultraviolet, visible, or near infrared light. It can be used with a wide range of targets, including nonmetallic objects, rocks, rain, chemical compounds, aerosols, clouds and even single molecules. For agriculture application, LiDAR can describe the shape and color of crops in 3D, which gives better evaluation basic for related analysis. However, the difficulty of visualization is increased since the density of 3D point cloud.

It is very important for experts and public to access the remote sense information about agriculture easily, so they can take actions accordingly. Considering the wide availability of WWW, it is a good choice to deliver remote sense data through browsers. Currently, many image based remote sense data are available, but there is not many 3D LiDAR cloud online visualization implementations. In current HTML4 framework, web browser itself does not support 3D visualization

S. Zhou and Z. Wu (Eds.): ADMA 2012 Workshops, CCIS 387, pp. 143–151, 2013.
DOI: 10.1007/978-3-642-41629-3_13, © Springer-Verlag Berlin Heidelberg 2013

natively. Therefore 3D senses can only be created with plugins which is difficult to develop and esp. hard to popularization.

Therefore, HTML5 related technologies are studied in this paper for online visualization of 3D LiDAR point clouds on agriculture. The rest of paper is structured as follows. Section 2 introduces related work. Section 3 describes the methodology of online visualization. Implementation details are explained in Sect. 4 and experimental results are discussed in Sect. 5. Section 6 concludes whole paper and suggests for future studies.

2   Related Work

LiDAR (Light Detection and Ranging) is an optical remote sensing technology that can measure the distance to, or other properties of a target by illuminating the target with light, often using pulses from a laser. As the incensement of its accuracy and point density, LiDAR is now applied in many applications to generate 3D models and analysis objects. Amzajerdian et al. [1] designed a LiDAR based system for Navigation and Safe Landing on Planetary Bodies; Lin et al. [2] tried to create 3D city models automatically from LiDAR data. In agriculture, LiDAR can generate a topographical map of the farm fields [3] and reveals the DEM and growth status of crops [4,5]. The LiDAR data is playing an increasingly important role in agriculture production. It is essential to share the information through Internet in3D, so that more people can have access to the data that could guide them in the crop production processes.

Three types of frameworks are proposed to visualization 3D scene through Internet: thick client-server model, thin client-server model and Peer to Peer (P2P) model. In the thick client-server model, the user has to install a specific client program to access the 3D scenes supplied by the city model server. Google Earth is the representation of think client framework. Isenburg and Shewchuk [6] described a set of streaming tools that allow quick visualization of large amounts of LIDAR data in Google Earth. Calle et al. [7] proposed a 3D point cloud visualization framework based on open source client GLOB3. In a thin client-server framework, the browser is usually used for retrieving, presenting, and traversing information resources on the web. Since current HTML4 do not support for 3D scenes natively, plugins are developed to visualize 3D LiDAR point clouds. Fabric Engine [8] is developed to visualize vast sets of geographic data such as LiDAR data in a web browser. DIELMO [9] is working in a project to implement different plugins for the distribution and visualization of LiDAR data with standard browser. The third type is peer-to-peer (P2P) online 3D visualization architectures that partitions data transmission and visualization tasks or workloads among peers. Lynch et al. [10] presented a multimedia visualization tool for large-scale mobile LIDAR using P2P network. These frameworks require installation and management in client side. It is necessary to develop a framework which is based on the standard technologies of Internet and is supported by the mainstream browsers in default. Therefore, HTML5 based technologies are employed in this paper for online visualization of agriculture related LiDAR point clouds.

HTML5 is the 5th major revision of the core language of the World Wide Web: the Hypertext Markup Language [11]. It is starting to support online 3D content [12] and is used for online 3D visualization. Chen and Xu [13] designed a platform for multiplayer online games based on WebGL [14], an HTML5 extension for 3D visualization. WebGL is a JavaScript API based on Open GL ES 2.0 enables the creation of 3D content in the web browser without any plugins. Based on WebGL, XB PointStream [15] proposed a framework to simplify the streaming and rendering of point clouds allowing them to be rendered in a web page without any plug-ins. To further simplify the 3D visualization in the browser, Behr et al. [16] proposed a framework, X3DOM, to directly integrate 3D city models in X3D format as the DOM element of HTML pages. Based on X3DOM, Mao and Ban [17] visualized 3D city models with standard browser. In this paper, X3DOM is selected to create 3D point cloud and Websocket is used to transmit the LiDAR data.

3   Methodology

The proposed online agriculture LiDAR point cloud visualization framework is illustrated in Fig. 1. First, the LiDAR dataset is transmitted into user client browser with Websocket with JavaScript method. Then the received point data is converted into DOM object in X3D format and directly updated in the HTML content. Finally, X3DOM will automatically visualize the 3D scenes in the created DOM objects based on WebGL.

The proposed framework is convenient to implement and can be extended according to different requirements. In the user client, if WebGL supported browser (such as Chrome, Firefox, Safari and Firefox mobile for android) is installed, the 3D point clouds will be visualized by the proposed framework without any plugin. In the proposed framework, LiDAR data, DEM data and 3D city models (CityGML) are stored in the database or file systems. Then the different datasets, such as the xyz file for LiDAR data, are parsed by java program and converted into Java objects. After model processing such as selection, simplification or generalization, the 3D scenes are created and encoded in X3D nodes. In the next step, these nodes are transmitted to the user Web browser using Websocket method. Finally, 3D scenes of the city models are presented to the user through the Web browser by X3DOM. The user can also interact with the 3D models and send feedbacks to the server through X3DOM. Based on the user requests, server can update the 3D models and create the new X3D nodes, which implement the interactions between the user and 3D java objects.

This framework has four main features. First, it is easy to extend the Java objects in server side for other purposes. We can easily get all 3D model information from java objects, process the data according to the request, make the analysis and generate the required output that could be in any format such as KML, O3D etc, in addition to X3D nodes.

Second, Websocket is used for the data transmission between the server and client which creates a better user experience compared with HTTP by reducing

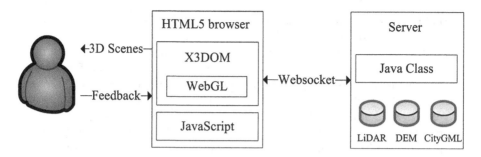

Fig. 1. HTML5 based online agriculture visualization framework.

the overhead and supplying full duplex communication. However, directly transmit the X3D scenes of a large area may lead to a longer response time. Therefore, the 3D models should be generalized and transformed to certain maxed LoD representation structures to reduce the data volume.

Third, developers do not need to deal with the different 3D visualization technologies or plugins in the Web browsers. X3DOM supports the standard visualization and interaction with X3D elements in the Web browsers. Along with the improvement of X3DOM, more features of X3D will be implemented and the 3D scenes generated in our framework will be accessible from increasing numbers of Web browsers.

Fourth, the interactions between user and the 3D scenes are enhanced because, on one hand, X3D itself support many sensors to detect user actions such as touch and movement, and on the other hand, the current 3D scenes can be updated dynamically from server according to these user requests.

The proposed framework is therefore efficient to integrate 3D data from different sources, to extend the framework for different purposes, to update 3D scenes dynamically and to support the interactions with user requests. A prototype of the framework is implemented to show these features. Next, we will introduce implementation details of the proposed framework.

4   Implementation

4.1   Agriculture LiDAR Dataset

Two types of LiDAR data on agriculture are studied in this paper. The first type was acquired by the helicopter-based FLI-MAP 400 system from John Chance Land Surveys, Inc over Enschede in Netherlands in 2006. The FLI-MAP 400 system is a LiDAR sensor integrated with an additional line scan camera, which is able to provide true color values to each laser point (Red, Green and Blue attributes can be associated to the laser data points). Therefore, point cloud and image data are consistently integrated in this dataset. The test area covers the area of $153 \times 278\,\mathrm{m}$ in ground but with irregular shape, and contains 1,197,686 LiDAR points integrated with RGB information.

The second type full waveform data were collected with the Riegl LMS-Q560 scanner. Table 1 shows the specifications of the instruments.

Table 1. Specifications of Riegl LMS-Q560 scanner.

Label	Location
Sensor	Riegl LMS-Q560
Scanning mechanism	Rotating polygon mirror
Wavelength (nm)	1550
Pulse width @ half maximum (cm; ns)	60; 4
Standard deviation of pulse (cm; ns)	25; 1.7
Recording of emitted pulse	Yes
Vertical sampling (m)	0.149855
Scan angle (u)	22.5
Effective measurement rate (kHz)	45
Beam divergence (mrad)	0.5
Flying height (m)	400
Size of footprint (cm)	20

These high resolution LiDAR point datasets on agriculture fields can well represent the features of crops in 3D, which is essential to analyses growth status and predict the productions. The first step of analysis the datasets is to make them available for the researchers and public. To increase the accessibility of the valuable data, online 3D visualization is necessary.

4.2   Data Transmission with Websocket

Different from HTTP (Hypertext Transfer Protocol, used for web page transmission), Websocket is based on TCP (Transmission Control Protocol) and can provide bi-directional, full-duplex communications channels. Websocket can reduce the overhead and increase the transmission speed effectively compared with HTTP. For example, the length of request and response header in HTTP usually is hundreds to thousands bytes, while it is only two bytes in Websocket. Also, the communication in Websocket is complete duplex so that browser and server can send data to each other anytime, while HTTP implements the half-duplex polling solution that only supports response-request model. Therefore, HTML5 Websockets can provide 500:1 or depending on the size of the HTTP headerseven a 1000:1 reduction in unnecessary HTTP header traffic and 3:1 reduction in latency [18].

In this paper, we select Jetty [19] as Websocket server. The key feature of the Jetty Websocket implementation is that it is fully integrated into the Jetty HTTP server and servlet container, so a Servlet or Handler can process and accept a request to upgrade a HTTP connection to a Websocket connection. To use Jetty Websocket framework, we first need to create a handler class to deal with socket communication. This class will manage all connected browsers, process the received requests and send back the required data. Then a server should be created in a port, combine the predefined handler and start the server to listen the connection from client browsers.

In Jetty, we only need to extent WebSocketHandler class and rewrite certain functions in to fulfill the user request. The first one is onOpen() function. It will be invoked when the connection con is established for the first time and it stores the new Websocket connection con in a connection pool. The second is onClose() function. It will remove corresponding connection con from the connection pool. The third one is onMessage() that is invoked when the server receives a message from user browser. If the message is request for the data, it will send the point cloud data to the client by conn.sendMessage(data) function. In this paper, data points is represented as P=(x,y,z,c) in which x, y and z are coordinate of the point and c is color (c is rgb in FLI-MAP 400 data and c is grey value in Riegl LMS-Q560 scanner data). All different points are joint as string and send to the user browser.

In user browser side, Websocket connection can be created with JavaScript by just specific the location such as "ws://hostname:port/". Then we can send and received text message to or from the server. The received point data with color information will be used to created X3D scenes with X3DOM.

4.3   Visualization with X3DOM

After receiving 3D point data from server, X3D scenes are created and visualized through browser with X3DOM. The visualization process is composed by updating DOM in HTML. In X3D specification, PointSet is suitable for the visualization of point cloud data. PointSet node in X3D specifies a set of 3D points, in the local coordinate system, with associated colors at each point. It is defined as follows in VRML format.

```
PointSet : X3DGeometryNode {
 SFNode [in,out] color    NULL [X3DColorNode]
 SFNode [in,out] coord    NULL [X3DCoordinateNode]
 SFNode [in,out] metadata NULL [X3DMetadataObject]
}
```

The coord field specifies Coordinate and the color defines the color for each point in order. The code in JavaScript to generate PointSet in X3D is listed as follows.

```
var s = document.createElement('Shape');
var b = document.createElement('PointSet');
var crd = document.createElement('Coordinate');
c.setAttribute("point", coord);
var clr = document.createElement(Color);
c.setAttribute("color", color);
b.appendChild(clr);
b.appendChild(crd);
s.appendChild(b);
```

```
var ot = document.getElementById('x3d_root');
ot.appendChild(s);
```

The code will generate a segment of X3D file in the web page and append it to the X3D root node as follows:

```
<x3d_root>

  <Shape>
    <PointSet>
      <Color color=''/>
      <Coordinate point=''/>
    </PointSet>
  </Shape>
</x3d_root>
```

5   Experimental Results

The proposed framework is implemented in LAN network environment. The server is Tomcat 7 running on a PC with Intel Core2 Quad 2.33 GHz CPU, 3.00 GB RAM and Microsoft Window XP SP3. The browser is Mozilla Nightly 13.0 (Test version of Firefox) running on a laptop with Intel Core2 Duo 2.40 GHz CPU, 2 GB RAM

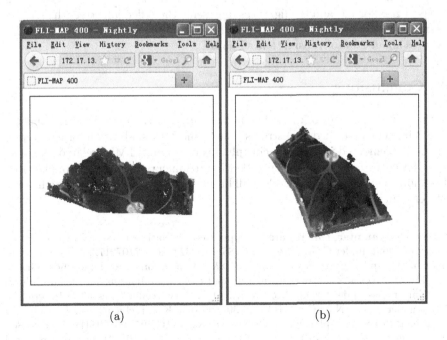

(a)   (b)

Fig. 2. Visualization results of FLI-MAP 400 dataset.

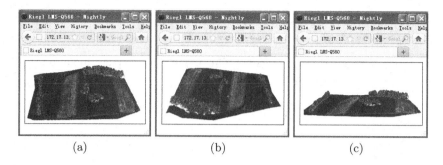

Fig. 3. Visualization results of Riegl LMS-Q560 dataset.

and Microsoft Window XP SP3. The development platform is Eclipse 3.7.1 running on the server.

The test datasets of the types of LiDAR point cloud are stored in a txt file and are loaded into the server as soon as it starts. When a client browser connected to the server, it will automatically creates a Websocket for the point data and visualizes with X3DOM.

Figure 2 demonstrates the visualization results of FLI-MAP 400 dataset in different viewpoint. This dataset contains about 115815 points. From the visualization results, we can clearly see the roads, trees and grass. The total load and visualization time for this dataset is around 10 s since the data is not compressed.

Figure 3 illustrates the visualization results of Riegl LMS-Q560 dataset. This one is in gray format and the lightness represents the reflection strength. The different types of crops can be recognized. The total number of point is 187310 which requires about 20 s to transmit and visualize.

6   Conclusions

In this paper, a framework of online agriculture LiDAR point dataset visualization is implemented. It demonstrates that online 3D visualization for agriculture can be implemented without specific plugins based on HTML5 related technologies. However, the performance is rather limited in current mainstream platform. Therefore, simplification methods shall be developed to increase the visualization efficiency in the future.

Acknowledgments. This research is supported by National Natural Science Foundation of China under Grants No. 41201486, 61103229 and 71072172, Jiangsu Provincial Colleges and Universities Outstanding S&T Innovation Team Fund under Grants No. 2011013, Key Project of Natural Science Research in Jiangsu Provincial Colleges and Universities under Grants No. 12KJA520001, Key Technologies R&D Program of China under Grants No. SQ2013BAJY4130, National Key Technologies R&D sub Program in 12th five-year-plan under Grants No. SQ2011GX07E03990, the Natural Science Foundation of Jiangsu Province of China under Grant BK2012863, International

S&T Cooperation Program of China under Grants No. 2011DFA12910, and Transformation Fund for Agricultural Science and Technology Achievements under Grants No. 2011GB2C100024. Jiangsu Province demonstration project of Internet of things (Yurun Group).

References

1. Farzin, A., Pierrottet, D.F., Petway, L.B., Hines, G.D., Roback, V.E.: Lidar systems for precision navigation and safe landing on planetary bodies. Technical report, Langel Research Center, NASA (2011)
2. Lin, H., Jing, H., Zhang, L.: Fast reconstruction of three dimensional city model based on airborne lidar. In: The International Archives of the Photogrammetry, Remote Sensing and Spatial Information Sciences, vol. XXXVII, Part B1, Beijing (2008)
3. Parrish, C.E., Nowak, R.D.: Journal of surveying engineering. Data Min. Knowl. Discov. **135**(2), 72–82 (2009)
4. Schulz, W.H.: Landslide susceptibility revealed by lidar imagery and historical records, seattle, washington. Eng. Geol. **89**, 67–87 (2007)
5. Zheng, G., Moskal, L.M.: Retrieving leaf area index (lai) using remote sensing: theories, methods and sensors. Sensors **9**, 2719–2745 (2009)
6. Isenburg, M., Shewchuk, J.: Visualizing lidar in google earth. In: 17th International Conference on Geoinformatics, pp. 1–4, August 2009
7. Calle, M.D.L., Gómez-Deck, D., Koehler, O., Pulido, F.: Point cloud visualization in an open source 3d glob3. In: ISPRS Archives, vol. XXXVIII-5/W16 (2011)
8. Dempsey, C.: Fabric engine: 3d lidar rendering via web browsers. http://gislounge.com/fabric-engine-3d-lidar-rendering-via-web-browsers/
9. dielmo. http://www.dielmo.com/eng/
10. Lynch, J.D., Chen, X., Hui, R.B.: A multimedia approach to visualize and interact with large scale mobile lidar data. In: Proceedings of the International Conference on Multimedia (MM '10), pp. 1689–1692. ACM, New York (2010)
11. W3C: Html5. http://dev.w3.org/html5/spec/Overview.html (2012)
12. Jianping, Y., Jie, Z.: Towards html 5 and interactive 3d graphics. In: 2010 International Conference on Educational and Information Technology (ICEIT), Chongqing China, pp. 522–527, September 2010
13. Chen, B., Xu, Z.: A framework for browser-based multiplayer online games using webgl and websocket. In: International Conference on Multimedia Technology (ICMT), Hangzhou China, pp. 471–474, September 2011
14. WebGL: Webgl - opengl es 2.0 for the web. http://www.khronos.org/webgl/ (2011)
15. Xb pointstream. http://zenit.senecac.on.ca/wiki/index.php/XBPointStream (2012)
16. Behr, J., Eschler, P., Jung, Y., Zöllner, M.: X3dom: a dom-based html5/x3d integration model. In: Web3D '09, Darmstadt, Germany (2009)
17. Mao, B., Ban, Y.: Online visualisation of a 3d city model using citygml and x3dom. Cartographica **46**(2), 109–114 (2011)
18. Lubbers, P., Greco, F.: Html5 web sockets: a quantum leap in scalability for the web. http://peterlubbers.sys-con.com/node/1315473
19. Jetty. http://jetty.codehaus.org/jetty/

Large-Screen Intelligent Display Platform
for Run-Time Data Visualization

Qingguo Wang[⊠]

Jiangsu Electric Power Information Technology Co. Ltd, Nanjing 210018, China
wang_qingguo@sohu.com

Abstract. Large-screen intelligent display platform is based on the modern
computer technology such as Internet technology, virtualization technology,
and etc. Also, it is the extension of SG - ERP. Run-Time data visualization can
help understand the relationships of different datas and improve the user
experience. In this work, a large-screen intelligent platform for run-time data
visualization is presented which can also provide guidance for other operation
monitoring center and applications of large-screen display technology.

Keywords: SG - ERP · Data visualization · User experience

1 Background

The new five-year informatization construction pointed out the direction of future
development for State Grid in China. And the goal is 'broader service coverage,
deeper integration and intelligence, higher safety, better interactive and visualization'.
And its Jiangsu subsidiary puts forward a series of slogans to improve the quality of
services, such as 'strive to create two top class pioneers', 'create a new situation in the
leading development', 'digitization supports intensive development' and 'informatization supports process control'.

Visualization [1–3] is a kind of technology to abstract things or processes into
graphic representation. Large-screen intelligent display platform for run-time data
visualization is based on the modern database technology [4], Internet technology [5],
GIS technology [6], modern communication technology [7], virtualization technology
[8], video compression [9] and transmission technology [10], absorbing informatics,
aesthetics, psychology etc. It is an effective way to solve the problem of 'Information
Island' [11].

2 Framework of Large-Screen Intelligent Display Platform
for Run-Time Data Visualization

The framework of large-screen intelligent display platform for run-time data visualization is stable now. As the chart shows, the system frame diagram has three layer

S. Zhou and Z. Wu (Eds.): ADMA 2012 Workshops, CCIS 387, pp. 152–158, 2013.
DOI: 10.1007/978-3-642-41629-3_14, © Springer-Verlag Berlin Heidelberg 2013

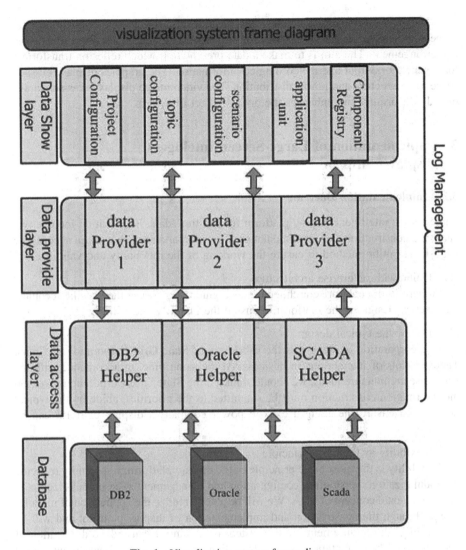

Fig. 1. Visualization system frame diagram

structures: data show layer, data provide layer, data access layer which can be seen in Fig. 1.

Data access layer includes four modules, the data object, data access object interface, data access object and data access object factory. The agile design improves the system's portability and scalability.

Data provide layer can classify the data by its sources, then reveals the global appearance and regional distribution of infrastructure facilities. It can realize the deep function of the system. The data mainly comes from the map of the province in two-dimensional. Because of the high resolution of large-screen, system needs to reset the scale and rendering to meet the large screen.

Data show layer mainly includes four basic levels: component, application unit, scene and theme. Five actions include registration, configuration, combination, layout and arrangement. The aim is to make a data presentation which relies on transformer load, generator output and meteorological information. The graphics display relies on space truss structure and power distribution. The video image display takes advantage of 3-d GIS display and enhances the display effect of system.

3   Implementation of Large-Screen Intelligent Display Platform

3.1   Implementation Doctrine

Large-screen intelligent display platform for run-time data visualization include two parts: the construction and implementation. The two parts need to work in accordance with the scientific method to ensure the working of the rationality and validity.

(1)   Guide with enterprise architecture

Adhere to the enterprise architecture as a guide. Try to use the specific technical architecture design of the platform to ensure the integrity and validity [12].

(2)   Follow the typical design

The corporation of Jiangsu is the subsidiary of State Grid Corporation of China. There are lots of interactions in business. At the same time, in accordance with the enterprise architecture ideas, we should think on the State Grid Corporation level. So the information construction must be submitted to the priorities, abide by the typical design, and ensure the Jiangsu electric power company's display platform can be energetic.

(3)   Flexibility as the basic principle

Flexibility is the most basic principle of the display platform. It is mainly reflected in modularization, versioning, configuration and management. The modularization is the fundamental concept of SOA. We will be able to realize the application of modular design through the construction and implementation of display platform and will be able to meet the requirements of the system reuse, and it can reduce duplication of construction. The connotation of modularization here is not only about the integrated interface component in SOA that we usually speak, but also combined with the display modularization, function modularization and so on. It is used to satisfy the flexibility in design and development [13].

It is provided a convenient for the operation of system through the simplification of the configuration. On the one hand, operation and maintenance personnel can use the simple configuration to realize the user's need; On the other hand, management personnel can inquire the information by the configuration information and it provides a condition to analyze the use of the application.

3.2   Implementation Process

In recent years, with the development of large-screen intelligent integrated display platform for run-time data visualization, the company has gained rich experience in project, the main experience are listed as followed:

(1)   Full investigation, promote demand

The company extracted technology personnel to establish project team and carried out the technical research and argumentation. On the one hand, we made numerous researches about the related hardware equipment of large screen in domestic and international, then we became familiar with the hardware environment and technical parameters. On the other hand, we made a research about the application of large screen in the industry inside and outside, and became in-depth understand the current trend of large screen. According to the research results, we could establish simulation environment and carry out a lot of testing work, and then grope a complete set reasonable scientific work ideas and technical route.

The project team attaches great importance to the business requirements analysis. The business department often presents simple and abstract requirement description, so the project team not only have to do from the user's core business, but also have to deeply understand the user's management concept and ideas to promote the management level.

(2)   Technological innovation, keeping improving

Both technological innovation and business expansion request the project team to keep learning, pick up advanced technologies, try new techniques and keep improving. In the development process of batches of intelligent display projects, the project team breaking the routine, trying new techniques, solving problems, has completed a number of applications for the first time and achieved self-improvement and self-transcendence [14].

(3)   Fast response, quick implement

Consideration of the large-screen intelligent integrated display platform for run-time data visualization can be able to response users' demands in time, so the visual configuration tool has been designed and developed. This tool based on a principle of "what you see is what you get", and makes the graphical interface easily to use. After a short training, every management consultant can use the configuration tool to complete business scene configuration.

Application architecture of visual configuration tool can be summarized into "three, four, five", which means facing three system role——developer, management consultant and terminal user, four basic levels——subassembly, application unit, scene and theme, completing five action——register, configuration, combination, layout, arrangement [15].

(4)   Scientific management, quality control

Since the large-screen visual intelligent display for run-time data visualization project started up, the project team has been strengthening scientific management, strictly controlling quality, and has put forward the slogan of the zero error. The team pays attention to quality control and minutiae, focusing on project plan, business needs, system development, system test, test run and other processes, in order to guarantee the quality of the whole hand-over system. The large-screen visual intelligent integrated display system of electric power dispatching control center of Jiangsu Company has run stably for more than one year, that system of material control center for almost one year, and all the using systems never make any serious mistake.

(5)   Personnel training, team building

At the beginning of the large-screen visual intelligent integrated display project startup, lack of talents, techniques and information. Project group cultivate a number of technical personnel and business experts. In the project development and implementation process, by collecting corporate wisdom, learning from others, independent innovation, following experienced staffs and so on. Excellent talents have formed a good team. In the project construction process, company has created a skillful, strong-willed research and development team, which possessed persistent pursuit, responsibility of the high level of discipline, tacit cooperation, continuous innovation and first-rate service, and has formed a firm-specific spirit of "large screen" and revealed the company's brand image.

4   Improvement of Large-Screen Intelligent Integrated Display Platform for Run-Time Data Visualization

With the development of technology, large-screen intelligent integrated display platform for run-time data visualization still need further improvement. The main targets are shown below:

4.1   Improve Visual Technology and Configuration

As the size of the display screen is more and more big, the display resolution becomes more and more high, business display content is more and more fine, display system visualization desktop configuration has become a comprehensive direction. Display platform has upgraded from the initial form type configuration to Diagram form configuration. We expect to build more practical and easier visual configuration tools to use and make the business scene configuration more agile and convenient.

4.2   Enhance the Interactivity of Show Scene

The current large-screen intelligent integrated display platform for run-time data visualization take more consideration of flattening panoramic display, but take little focus on content of traceability and interaction. How to improve interactive of display scene has become a new direction of the display platform. Depends on the current situation, We will focus on the improvement of dynamic business ability, and enhance the interactivity of application unit and scene.

4.3   Enhance the User Experience

The current large-screen intelligent integrated display platform for run-time data visualization takes more consideration to show business scene, but lack of interaction with users. Enhancing the user experience has become one of the development goals. Next step, we will focus on how to use the touch screen and body feeling interaction technology to improve display platform, make a demonstration and upgrade the large-screen intelligent display platform for run-time data visualization to enhance user experience.

5   Conclusion

It's a long-term task to build and improve the intelligent display platform. In the following work, it must continue to strengthen the platform team construction and establish normalization work mechanism, and provide support for the information construction in order.

Acknowledgments. This research is supported by International S&T Cooperation Program of China under Grants No. 2011DFA12910.

References

1. Moore, J.H., Cowper-Sallari, R., Hill, D.P., Hibberd, P., Madan, J.: Human microbiome visualization using 3d technology. In: Pacific Symposium on Biocomputing, pp. 154–164 (2011)
2. Wright, H., Crompton, R.H., Kharche, S., Wenisch, P.: Steering and visualization: enabling technologies for computational science. Future Gener. Comput. Syst. **26**(3), 506–513 (2010)
3. Huffman, J., Forsberg, A., Loomis, A., Head, J., Dickson, J., Fassett, C.: Integrating advanced visualization technology into the planetary Geoscience workflow. Planet. Space Sci. **59**(11), 1273–1279 (2011)
4. Mitrea, P., Deak, C.: Online diagnosis e-health system for all, based on advanced web accessible database technologies. Int. J. Knowl. Web Intell. **2**(1), 64–86 (2011)
5. Hiroi, K., Yamanouchi, M., Sunahara, H.: A proposal of disaster information system based on the Internet technologies In: IEEE - SENSORS, pp. 1848–1853 (2010)
6. Ahmad, S.Z, Abdullah, S.L.S., Rosmani, A.F., et al.: Integrating GIS technology and 3D animation in an event information system. In: International Conference on Science and Social Research – CSSR (2010)
7. Parikh, P.P., Kanabar, M.G., Sidhu, T.S.: Opportunities and challenges of wireless communication technologies for smart grid applications. In: Power Engineering Society, IEEE General Meeting – PES, pp. 1–7 (2010)
8. Kim, I., Kim, T., Ikeom, Y.: NHVM: design and implementation of linux server virtual machine using hybrid virtualization technology. In: Computational Science and Its Applications – ICCSA (2010)

9. Murakami, T.: Technical trends of next generation ultra high definition video-compression, transmission technologies and their standardizations. In: IEEE International Symposium on Broadband Multimedia Systems and Broadcasting - BMSB (2010)

10. Seki, T., Nishimori, K., Hiraga, K., Nishikawa, K.: Experimental evaluation of high speed parallel data transmission technology for wireless repeater system. In: IEEE Radio and Wireless Symposium Formerly - RWS Formerly RAWCON (2010)

11. Yu, X., Li, P., Li, S.: Research on data exchange between heterogeneous data in logistics information system. In: International Conference on Communication System Networks and Applications – ICCSNA, (2010)

12. Shen, G., Lv, J., Di, F., She, D., Sun, P., Zhao, L.: Research and application of network visualization display technology. Electr. Power Inf. **04**, 59–62 (2009)

13. Overbye, T.J., Wiegmann, D.A., Thomas, R.J.: Visualization of power systems: Final report, PSERC Publication 17-30, New York (2005)

14. Han, Z., Lv, J., Qiu, J.: Scientific computing visualization and its application prospect in power system. Power Syst. Technol. **07**, 22–27 (1996)

15. Sun, Y., Overbye, T.J.: Visualizations for power system contingency analysis data. IEEE Trans. Power Syst. **19**(4), 1859–1866 (2004)

A Pork Traceability Framework Based on Internet of Things

Baocai Xu, Jingjun Li$^{(\boxtimes)}$, and Yun Wang

Jiangsu Yurun Food Industry Group Co., Ltd, Nanjing, China
2003jjli@163.com

Abstract. It is essential to trace the source of food for security reasons. In this paper, a traceability framework is proposed and implemented for pork industry. We apply the technologies from Internet of Things such as remote monitoring, sensor network, and data mining to create a chain covering the whole process of pork production. The traceability information can also be accessed through the Internet and any costumer could find out where the pork he/she is buying comes from and how it comes. This paper is mainly focus on the implementation side of the traceability and how to preserve related information along the production process. Our experiments show that the proposed framework can implement the pork traceability efficiently.

Keywords: Traceability · Internet of things · 2D barcode · Data mining

1 Introduction

Food Traceability System covers production, inspection, supervision and consumption. From the end of the 20th century to the early 21st century, the European Union, Japan and United States developed frameworks to implementation the traceability of domestic animals and animal products. These countries apply food safety legislation or government regulation to force farms to adopt food traceability system, which includes Forage breeding, environment management, inspection and quarantine, slaughtering and processing storage, transportation and sales.

Recent years, China also realizes the importance of food safety traceability. The government has started the establishment of livestock production safety regulations. Some enterprises have also developed their animal traceability information system. However, only a limited number of food make full use of the tagging technology to its certain process, and these existing implementations are mainly based on human recording which will increase the cost of production. Therefore, Internet of Things should be applied into the food traceability system.

Technologies of Internet of Things provide efficient tools for information gathering, data communication and sharing from different resources automatically. Traceability System can make consumers understand the production and circulation process, and increase consumers' faith to the food itself. Therefore, it is necessary to apply Internet of Things to implement food traceability.

S. Zhou and Z. Wu (Eds.): ADMA 2012 Workshops, CCIS 387, pp. 159–166, 2013.
DOI: 10.1007/978-3-642-41629-3_15, © Springer-Verlag Berlin Heidelberg 2013

In this paper, a pork traceability framework is proposed based on Internet of Things. The rest of paper organized as follows: Related work is introduced in Sect. 2. Section 3 describes the proposed system. Section 4 gives the implementation and Sect. 5 concludes the whole paper.

2 Related Work

Recently, many countries are concerned about their safety of the food supply chain, and have established application system which includes animal husbandry industry safety regulations, hazard analysis, critical control point in animal products processing. A series of traceability application system are applied to supervise food supply chain, which leads to a quality and safety traceability system [2].

The food traceability system use the environmental monitoring, climate simulation, satellite tracking, imaging analysis technology to ensure the quality and safety of the food such as pork [1].

The EU's animal product traceability system is mainly used in the production of cattle and circulation field. The traceability system uses the unified central database to do information management. Japan comprehensively introduced the traceability system in beef cattle production and supply to cope with mad cow disease in 2001. And in June 2002, the system was generalized to the tracking system of pork and chicken meat industry, etc. The United States established PQA (pork quality assurance system) which based on the HACCP, focusing on the detection of the key control points.

Traceability system with large individual animal identification technology aims to safeguard animal health and to ensure human security, and integrates existing technology, such as iris recognition technology and DNA fingerprinting technique to establish the iris database, DNA fingerprint database and the corresponding large animal iris recognition system [4].

China's pig individual identification technology has just begun. The first set of indicators "Safety of Factory Pork Production Traceability Digital Systems" use ear tags as identification of feeding and management, inspection and quarantine, drug residues, prohibited additives traceability, but have not yet made information collection and identification technology research in other field, such as the livestock and poultry breeding, slaughter and processing of livestock, poultry products, storage, transportation and sale of the entire food safety information, and also do not complete the traceability system of the production and distribution of livestock and poultry meat.

3 Traceability Framework

This article aims to study the key technology of fresh pork safety traceability information system, and bases on fresh pork production and circulation of the whole supply chain (breeding, transportation, slaughter and products processing, storage, transportation, sales). The system combines with RFID automatic identification technology, sensor technology, ZigBee wireless communication technology, GPS/temperature detection technology, and based on cloud computing distributed data storage and processing technology, data mining and knowledge discovery technology and artificial intelligence technology, network technology, global positioning technology, web development technology, network query and search techniques etc. [5]. It collects information from breeding, transportation, slaughtering and processing to provide a series of services, such as information, product tracking, departments supervision, industry early warning, complaints rights and index analysis for the government, consumer and enterprise (Fig. 1).

The details of this research are listed as follows:

(1) Information tracking and collection system

At the stage of breeding in the farm, the electronic ear tag which carries the RFID chip will be embedded into newborn pig's ears. It will give a unique global ID to each pig, recording the information of feeding, feed, medicine and epidemic prevention. The sensor and wireless transmission technology on piggery will collect environment information, and store in the database through the combination of node arrangement [6].

At the stage of slaughter and processing, the reusable wireless RFID card will be hanging on every piece of pork, achieving effective conversion of individual identification information. The unified two-dimensional bar code is used in meat resin, visceral, head and leather unified code at the time of partitioning. The check and review information will be loaded onto the database, corresponding with code.

At the stage of transportation, the information of transportation and quality inspection will be conveying to database by wireless transmission before the slaughter. After the slaughter, use GPS/temperature detection technology, wireless sensor technology, electronic map and wireless transmission technology to establish an open supervision platform [7]. Not only effectively realize the refrigerated truck

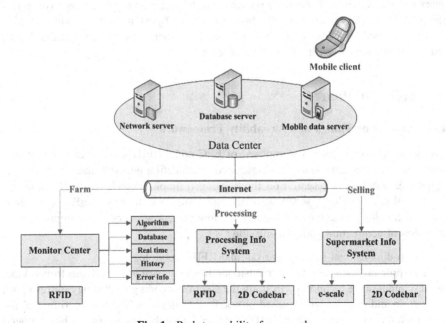

Fig. 1. Pork traceability framework

resources location tracking, but also realize the temperature data acquisition transmission, recording and transfinite alarm in the refrigerated car and fresh-preserved storehouse. It is offering a solution for the shipping and tracking.

At the stage of distribution and sales, the sensors collect storage, transportation, purchase, sellers, early warning and sales information. The reader reads information, use the supermarket internal one-dimensional bar code to replace the factory bar code, and put this information into the database server. Finally, it will be on sale.

(2) Data storage and process system

Data storage and processing is the core of the traceability system. Considering the actual needs of the company, such as the large scale of production, storage and transportation network, multiple attribute data, real-time update and frequently change, it is necessary to build a hierarchical distributed storage and integrated system to ensure the integrity of the data storage and high efficiency. The company decides to establish data model for the integrated data through the data mining, multi-dimensional analysis, service polymerization, data recommended and middleware technology to carry out data analysis and processing. The aim is to make further optimization of traceability system and improve the efficiency of traceability and feasibility and set up expert system that based on artificial intelligence. The company will establish the process control and early warning system.

(3) Quality and safety traceability service platform

After the research of cloud computing, the company builds a platform which based on net and web service technology B/S network, supporting the remote, multi-user, terminal equipment of the visual access rights and complaints. To provide users with industry early warning mechanism and department supervision channel [8]. Set up index release system and regularly present industry analysis report to the government and the public. The research results show the safety of pork and pork products, which satisfy the need of consumers and increase consumer trust of the manufacturers, promoting industry overall safety control level.

4 Implementation

4.1 Internet of Things in Traceability Framework

Internet of Things is mainly composed of EPC code, RFID tags, reader, neural network software (Savant), sensors, wireless communication network and database. The project team conducts research on the existing technology such as RFID automatic identification technology, sensor network and testing technology, ZigBee communication technology to achieve the goals for low cost, low energy consumption, high precision of acquisition and transmission.

(1) Two-dimensional barcode ear tag and RFID technology

The project team use two-dimensional barcode earmarks and electronic label (RFID) middleware technology to study the electronic coding technology of fresh pork production and circulation, standard protocol and interface. The aim is to realize the information interaction between RFID reader and traceability system and build related coding traceability database that based on electronic tags middleware technology [9]. The key point of research is to research integrated data filtering and RFID event manager with pretreatment system. It also studies the structure of electronic label middleware and RFID information server which has multi-function interface.

(2)   Effective information conversion for Individual identification

This project will research a kind of recycling RFID identification card that can record the slaughter and inspection process information, and at the end of the slaughter and processing, the information will be effectively converted to two-dimensional barcode for each piece of pork, achieving the goal for seamless connection and conversion of individual identification.

(3)   Sensor network and testing method

The focal point of the sensor is about the distributed information processing technology, embedded computer technology, intelligent sensor network node and sensor detection and control technology. The aim is to realize the information interaction among the database and information collection of piggery and transport.

(4)   Communication Based on ZigBee

ZigBee is a wireless communication technology that has many advantages. It is simple, low power consumption, low cost and high efficiency. Because its data rate is low and Network capacity is large, it is suitable for application in pig breeding, slaughtering of data transmission. At the same time, it can ensure the security of data.

4.2   Traceability Information Management

The data from the farms, processing factories, supermarkets and transport way has diverse features. It involves large volume, high redundancy, complex transfer and many replace links. It needs to study large amounts of data distributed storage and analysis, data classification and processing, data mining and recommend technology to provide technical support for traceability system [10]. On that basis, there is more to come. The artificial intelligence expert system and the monitoring and early warning technology should be studied to solve the quality and safety problems of pork in the supply chain in time.

(1)   Mass data distributed storage technology

It needs to store different characteristic value and different types of data, and also needs to manage and coordinate the underlying disk facilities for the construction of hierarchical distributed mass data storage system to ensure the high availability and efficiency.

(2)   Mass data analysis system

Using MapReduce programming framework to conduct distributed mass data analysis system. According to logic data flow in the data analysis application, the project team designs MapReduce processing procedure, which includes several steps, such as work master function, map, shuffle, combine and reduce. According to the actual demand, the project team needs to choose proper programming language.

(3) Mass data classification and processes

The project team uses association rules, sequential patterns, classification and clustering technology, data mining and knowledge discovery technology to extract effective and useful information from mass data. And the users can query relevant fresh pork analysis report through complicated aggregate queries.

(4) Artificial intelligence based expert system

The project team devotes to studying human-computer interaction technology, artificial neural network knowledge acquisition and learning mechanism, intelligent search, knowledge processing technology, and the intelligent control technology and intelligent signal processing technology. The intelligent system is implanted in ear tag that makes the sensing technology more digital and intelligent. The only fresh pork expert system will be built up to analysize the data from Yunrun and the industry intelligently. It can be used to simulate problem-solving strategies for fresh pork supply chain, and it will find out the hidden danger for enterprise and user to make intelligent decisions.

(5) Controlling and warning technology

Based on data processing and expert system, the early warning system and workflow link warning response condition should be constructed. It plays a key role in the process of early warning feedback. According to the different hazard analysis of fresh pork production and critical control point, the project establishes HACCP management component library which includes plan management module, key control point comparison module, record management module and deviation warning module.

The information construction of the grain industry is based on the specification of information technology. The main agency of the construction draw up the specifications, the coding of the exchange of information, the system architecture according to their own need because there is no uniform of guidance [10]. It will lead to the information construction system of all main agency cannot be combined organically and information silos. Grain circulation information will be collected only in the specific region and that information about exchange which is out of the region cannot be gathered. So it will lead to the information construction becoming difficult and information standards are not unified and various information systems are in their own way and increasing the phenomenon of repeating investment.

4.3 Individual and Pollutant Tracking

Make research of isotopic traceability technology, pollutant source technology, DNA individual identification and iris individual identification which mainly depend on existing technology, such as the Label traceability technology, origin traceability technology and food information tracking technology.

(1) Pig individual identification

To study the DNA and iris recognition model to establish permanent characteristics database which can be used as the basis for identification and coding. The system wants to integrate with existing iris recognition technology and method and make a breakthrough in extracting large animal individual iris feature and improving the recognition rate and robustness. In order to improve the whole system identification ability and the overall ability of traceability, the project team studies the iris and DNA identification information fusion algorithm, and puts forward an individual identification procedures and application guide for the pig from "table to base".

(2)   Origin and pollutant tracing

The aim is how to apply the traceability technology for radioactive nuclide, heavy metal and biological toxin pollution detection. According to these characteristics, such as environmental elements, production data of radionuclide and heavy metal pollutants of the isotopic composition, the system will find the origin pollutant source and practical isotope fingerprint. Based on reserved samples, the system makes use of radionuclide and heavy metals pollutants isotope that exist in the processing procedure to study pollutants traceability technology and practical isotope fingerprint [11]. The approach is to analysis the isotopic characteristics of biological toxins during the raw material production process, food processing, storage and transportation process, and finally establish a full traceability system of food contamination, and put forward the application guidelines and procedures for traceability.

5   Conclusion

Pork accounts for two-thirds of the meat eaten by Chinese consumers, but only half of it goes through slaughterhouses that are subject to inspection. Safety problem of fresh pork in China has been one of the major problems. The traceability system needs to be implemented during the overall process urgently. The whole system not only ensure the safe and reliable of the pork products, but also avoid the pork product from dead or ill pig, or illegally produced ones flow into the market while satisfying the needs of pork product whole life cycle management, and it can accomplish the whole process supervise of the pork products and make the consumers buy them with confidence and eat them with satisfaction.

With the development of information and communication technology, more and more technology will be applied in the pork safety traceability system. Not only use the animal identification technology, information technology and terminal technology, but also make great progress in some modern communication technology and analysis techniques. For example, microwave radar can be used in animal tracking, general packet radio service can be used for traceability information of commonly used inquires. In a word, with the development of the society, new technology and equipment will be applied to the information traceability of pork-products to achieve more accurate tracking.

Acknowledgments. This research is supported by jiangsu science and technology support item under Grants No.BE2011398.

References

1. Xie, J.F., Lu, C.H., Li, B.M., Wang, L.F., Shi, Y., Xue, Q.K., Li, B.S.: Development of monitoring and traceability system for pork production. World Engineers Convention, Shanghai, China, pp. 2–6 (2004)
2. Hobbs, J.E., Von Bailey, D., Dickinson, D.L., Haghiri, M.: Traceability in the Canadian red meat sector: do consumers care? Can. J. Agric. Econ. **53**, 47–65 (2005)
3. Cao, J., Wu, Z., Mao, B., Zhang, Y.: Shilling attack detection utilizing semi-supervised learning method for collaborative recommender system. World Wide Web J. Internet Web Inf. Syst. (2012). doi:10.1007/s11280-012-0164-6
4. Arana, A., Sovet, B., Lasa, J.: Meat traceability using DNA markers: application to the beef industry. Meat Sci. **61**, 367–368 (2002)
5. Wu, Z., Mao, B., Cao, J.: MRGIR: open geographical information retrieval using MapReduce. In: The 19th International Conference on GeoInformatics (GeoInformatics 2011), Shanghai, China, June 2011
6. Buckley, J.: From RFID to the Internet of things—pervasive networked systems. In: European Commission, DG Information Society and Media, Networks and Communication Technologies Directorate, Brussels (2006)
7. Xiong, B.H.: A Comprehensive Technical Platform for Precision Feeding of Dairy Cattle, pp. 51–57. Chinese Agricultural Science &Technology Press, Beijing (2005). (in Chinese)
8. Cao, J., Wu, Z., Wang, Y., Yi, Z.: Hybrid collaborative filtering algorithm for bidirectional web service recommendation. Knowl. Inf. Syst. (2012). doi:10.1007/s10115-012-0562-1
9. Schmidt, O., Quilter, J.M., et al.: Inferring the origin and dietary history of beef from C, N and S stable ration analysis. Food Chem. **91**, 545–549 (2005)
10. Cunningham, E.P., Meghen, C.M.: Biological identification systems: genetic markers. Sci. Tech. Rev. **20**(2), 491–499 (2001)
11. Buitkamp, J., Ammer, H., Geldermenn, H.: DNA fingerprinting in domestic animals. Electrophoresis **12**, 169–174 (1991)

Author Index